Magnetic Functional Materials: Synthesis, Characterization and Application

Editors

Haiou Wang

Dexin Yang

MDPI • Basel • Beijing • Wuhan • Barcelona • Belgrade • Manchester • Tokyo • Cluj • Tianjin

Editors
Haiou Wang
Hangzhou Dianzi University
China

Dexin Yang
Hangzhou Dianzi University
China

Editorial Office
MDPI
St. Alban-Anlage 66
4052 Basel, Switzerland

This is a reprint of articles from the Special Issue published online in the open access journal *Materials* (ISSN 1996-1944) (available at: https://www.mdpi.com/journal/materials/special_issues/mag_mat).

For citation purposes, cite each article independently as indicated on the article page online and as indicated below:

LastName, A.A.; LastName, B.B.; LastName, C.C. Article Title. *Journal Name* **Year**, *Volume Number*, Page Range.

ISBN 978-3-0365-5467-9 (Hbk)
ISBN 978-3-0365-5468-6 (PDF)

Contents

About the Editors

Haiou Wang

Haiou Wang is an associate professor in the Department of Materials Science and Engineering at Hangzhou Dianzi University in China, as of 2018. Wang was awarded his Ph.D. degree in Materials Science and Engineering from the Nanjing University of Science and Technology and his bachelor's degree in Applied Physics from Jiangsu Normal University in China. After obtaining his Ph.D. degree, he joined Hangzhou Dianzi University in 2014. Over the last two decades, he dedicated all his efforts to the synthesis of magnetic functional materials and their applications to magnetic devices. To date, he has published over 60 peer-reviewed articles, including Applied Physics Letters; Europhysics Letters; Materials Letters; and Journal of Applied Physics.

Dexin Yang

Dexin Yang is currently an associate professor at Hangzhou Dianzi University and a visiting scholar at the College of Optical Science and Engineering, Zhejiang University. He received his B.Sc. in 2012 and Ph.D. in 2016 from the China University of Geosciences (Beijing). He was a visiting student at the University of Cambridge, UK (2014–2015). He was a postdoctoral research fellow at the same department at Zhejiang University from 2019 to 2021. His research concerns the roles of strain and elastic relaxation in functional materials, as well as the phase transition, ferroelasticity, and magnetic and optoelectronic properties of perovskite semiconductors. He has published over 20 peer-reviewed articles as first or corresponding author, including Nature Communications; Physical Review B; and Advanced Functional Materials.

Preface to "Magnetic Functional Materials: Synthesis, Characterization and Application"

The early magnetic materials were mainly silicon steel and ferrite. Since the 1960s, a series of high-performance magnetic functional materials, such as amorphous soft magnets, nanocrystalline soft magnets, and rare-earth permanent magnets, have appeared one after another. Driven by contemporary advanced science and technology, new properties and new phenomena of magnetic functional materials are emerging, and their fields are becoming wider and wider.

In the contemporary information society, energy, information, and materials are the important foundation of production, life, and high technology. Magnetic functional materials are widely used in energy, information, and materials science and technology. There are many kinds of magnetic functional materials, and their progress is rapid. Magnetic functional materials have attracted a great deal of attention regarding their applications. Magnetic behaviors are widespread in a variety of materials, such as metals, ceramics, organics, and emerging 2D materials. Applications of magnetic materials include memories, sensors, magnetic refrigeration, drug delivery, electrochemistry, environmental protection, energy storage, and more.

This Special Issue aims to publish original and review papers on new scientific and applied research and make great contributions to the discovery and understanding of magnetic functional materials and related syntheses, fundamentals, characterization, and applications. Of the twelve research articles submitted, nine were accepted for publication after the peer-review process, an acceptance rate of 75 percent. The published articles cover a range of topics and applications central to magnetic functional materials.

Finally, we would like to take this opportunity to express our most profound appreciation to the MDPI Book staff; the editorial team of the Materials journal, especially Mr. Felix Guo; the assistant editor of this Special Issue; the talented authors; and the hardworking and professional reviewers.

Haiou Wang and Dexin Yang
Editors

Editorial

Magnetic Functional Materials: Synthesis, Characterization and Application: A New Open Special Issue in Materials

Haiou Wang *, Yan Wang and Dexin Yang

College of Materials and Environmental Engineering, Hangzhou Dianzi University, Hangzhou 310018, China; 202040219@hdu.edu.cn (Y.W.); dy263@hdu.edu.cn (D.Y.)
* Correspondence: wanghaiou@hdu.edu.cn

Citation: Wang, H.; Wang, Y.; Yang, D. Magnetic Functional Materials: Synthesis, Characterization and Application: A New Open Special Issue in Materials. *Materials* **2022**, *15*, 2999. https://doi.org/10.3390/ma15092999

Received: 18 April 2022
Accepted: 19 April 2022
Published: 20 April 2022

Publisher's Note: MDPI stays neutral with regard to jurisdictional claims in published maps and institutional affiliations.

Magnetic Functional Materials: Synthesis, Characterization and Application is a new open Special Issue of *Materials*, which aims to publish original and review papers on new scientific and applied research, and make great contributions to the finding and understanding of magnetic functional materials and related synthesis, fundamentals, characterization, and applications.

The early magnetic materials were mainly silicon steel and ferrite. Since the 1960s, a series of high-performance magnetic functional materials such as amorphous soft magnets, nanocrystalline soft magnets and rare-earth permanent magnets have appeared one after another [1]. Driven by contemporary advanced science and technology, new properties and new phenomena of magnetic functional materials are emerging, and their fields are becoming wider and wider.

In the contemporary information society, energy, information, and materials are the important foundation of production, life, and high technology. Magnetic functional materials are widely used in energy, information, and materials science and technology. There are many kinds of magnetic functional materials, and their progress is rapid. Magnetic functional materials have attracted a great deal of attention regarding their applications. Magnetic behaviors are widespread in a variety of materials, such as metals, ceramics, organics, and emerging 2D materials. Applications of magnetic materials include memories, sensors, magnetic refrigeration, drug delivery, electrochemistry, environmental protection, energy storage, and more.

The research interest of the section *Magnetic Functional Materials: Synthesis, Characterization and Application* includes, but is not limited to, the following: permanent magnets; magnetic functional materials; magnetism in correlated electron systems; memories and sensors devices; magnetic refrigeration; environmental protection; and devices based on magnetic materials.

Funding: This work was supported by the National Natural Science Foundation of China (No. 11604067).

Conflicts of Interest: The authors declare no conflict of interest.

Short Biography of Authors

Haiou Wang is an associate professor in the Department of Materials Science and Engineering at Hangzhou Dianzi University in China as of 2018. Wang was awarded his Ph.D. degree in Materials Science and Engineering from Nanjing University of Science and Technology and his bachelor's degree in Applied Physics from Jiangsu Normal University in China. After obtaining his Ph.D. degree, he joined Hangzhou Dianzi University in 2014. Over the last two decades, he dedicated all his efforts to the synthesis of magnetic functional materials and their applications to magnetic devices. To date, he has published over 50 peer-reviewed articles.

Dexin Yang is currently an associate professor at Hangzhou Dianzi University, and a visiting scholar at the College of Optical Science and Engineering, Zhejiang University. He received his B.Sc. in 2012 and Ph.D. in 2016 from China University of Geosciences (Beijing). He was a visiting student at the University of Cambridge, UK (2014–2015). He was a postdoctoral research fellow at the same department at Zhejiang University from 2019 to 2021. His research concerns the roles of strain and elastic relaxation in functional materials, as well as the phase transition, ferroelasticity, and magnetic and optoelectronic properties of perovskite semiconductors. He has published over 20 peer-reviewed articles as first or corresponding author, including Nature Communications; Physical Review B; and Advanced Functional Materials.

Reference

1.   Coey, J.M.D. Magnetic Materials. *J. Alloy. Compd.* **2001**, *326*, 2–6. [CrossRef]

 *materials*

Article

Comparisons of Dy Utilization Efficiency by DyH$_x$ Grain Boundary Addition and Surface Diffusion Methods in Nd-Y-Fe-B Sintered Magnet

Shuai Guo [1,2,3], Shicong Liao [2,3], Xiaodong Fan [1,2,3,*], Guangfei Ding [1,2,3], Bo Zheng [1,2], Renjie Chen [1,2,3] and Aru Yan [1,2,3,*]

[1] CISRI & NIMTE Joint Innovation Center for Rare Earth Permanent Magnets, Ningbo Institute of Material Technology and Engineering, Chinese Academy of Sciences, Ningbo 315201, China
[2] Key Laboratory of Magnetic Materials and Devices, Ningbo Institute of Material Technology and Engineering, Chinese Academy of Sciences, Ningbo 315201, China
[3] University of Chinese Academy of Sciences, Beijing 100049, China
[*] Correspondences: fanxiaodong@nimte.ac.cn (X.F.); aruyan@nimte.ac.cn (A.Y.)

Abstract: Using the heavy rare earth Dy element to improve coercivity is the most common solution for hindering the reduction in magnetic properties in the Nd–Fe–B magnet, and the effective utilization of Dy has become the focus of research in industrial society. In this work, we investigated the influence of DyH$_x$ addition and diffusion methods on the microstructure, magnetic performance, and thermal stability of the Nd–Y–Fe–B magnet with a Y-rich core structure. The coercivity of the DyH$_x$ addition magnet increases from 9.45 kOe to 15.51 kOe when adding 1.6 wt.% DyH$_x$, while the DyH$_x$ diffusion magnet increases to 15.15 kOe. According to the analysis of the microstructure and elemental distribution, both Dy-rich shells were basically formed due to the diffusion process of Dy atoms. The Dy-rich shell in the DyH$_x$ addition magnet was similar with the original core–shell structure in the Nd–Y–Fe–B magnet. However, the distinct dual-shell structure consisting of a thinner Dy-rich shell and a Y-lean shell was constructed in the DyH$_x$ diffused magnet, contributing to the superior coercivity increment and Dy utilization efficiency. Furthermore, the remanence of the DyH$_x$ diffused magnet is up to 12.90 kG, which is better than that of the DyH$_x$ addition magnet (12.59 kG), due to fewer Dy atoms entering the 2:14:1 matrix grain to cause the antiferromagnetic coupling with Fe atoms. Additionally, the thermal stability of the DyH$_x$ diffusion magnet is also better than that of the DyH$_x$ addition magnet, owing to the elevated coercivity at room temperature, which expands the application range of the Nd–Y–Fe–B magnet to a certain extent.

Keywords: Dy utilization efficiency; DyH$_x$; addition method; diffusion method; Nd–Y–Fe–B magnet

Citation: Guo, S.; Liao, S.; Fan, X.; Ding, G.; Zheng, B.; Chen, R.; Yan, A. Comparisons of Dy Utilization Efficiency by DyH$_x$ Grain Boundary Addition and Surface Diffusion Methods in Nd-Y-Fe-B Sintered Magnet. *Materials* **2022**, *15*, 5964. https://doi.org/10.3390/ma15175964

Academic Editors: Haiou Wang and Dexin Yang

Received: 1 August 2022
Accepted: 24 August 2022
Published: 29 August 2022

Publisher's Note: MDPI stays neutral with regard to jurisdictional claims in published maps and institutional affiliations.

1. Introduction

Recently, the consumption of Nd–Fe–B-type magnets has rapidly increased all over the world because of their outstanding performance in electric motors, hybrid vehicles, wind generators, and electronic communication devices, etc. [1–10]. This leads to a large consumption of rare earth metal resources, such as Pr, Nd, Dy, or Tb, and a significant increase in the cost of magnets. Simultaneously, large amounts of the rare earth elements, such as Y, which possesses the merits of low cost and high abundance in the Earth's crust, have been left unused [11–14]. Thus, the substitution of the rare earth element Y for Nd [15–17] has attracted much attention in the permanent magnetic society.

However, the coercivity of the Nd–Y–Fe–B magnet significantly deteriorates after doping with Y due to the lower anisotropy field of Y$_2$Fe$_{14}$B (H_a = 26 kOe) compared with Nd$_2$Fe$_{14}$B (H_a = 73 kOe) [11,12]. The heavy rare earth elements Dy and Tb have been introduced directly to improve the coercivity of Nd–Fe–B magnets for the strong anisotropy fields of Dy$_2$Fe$_{14}$B (H_a = 150 kOe) and Tb$_2$Fe$_{14}$B (H_a = 220 kOe) [4] at room temperature.

Generally, a $(Dy/Tb, Nd)_2Fe_{14}B$ phase with enhanced anisotropy field will be formed on the matrix grain surface of the magnet, while Dy/Tb is doped into the magnet as the oxides [18,19], hydrides [20–22], fluorides [23,24] and so on, but the coercivity of the magnet is not enhanced as much as expected. Another method to introduce heavy rare earth Dy/Tb is the grain boundary diffusion process (GBDP) which is also the most important inventions in the last two decades for the rare earth (RE) permanent magnets industry [25–27]. It has been proposed to enhance the coercivity of Nd–Fe–B magnets using Dy/Tb diffused along the grain boundary into the interior and the matrix grains to construct a Dy/Tb-rich shell [28]. On the other hand, the remanence of Nd–Fe–B magnets will be heavily decreased due to the antiferromagnetic coupling among the Dy/Tb atoms and the Fe atoms [29,30]. Therefore, it is important to introduce heavy rare earth elements without damaging the other magnetic properties, that is to say, how to efficiently utilize heavy rare earth Dy/Tb in small quantities has become the theme of the current research.

In this work, the effect of two different ways of Dy introduction, i.e., DyH_x addition and DyH_x diffusion methods, on the magnetic properties, microstructures, and thermal stability of the Nd–Y–Fe–B magnet has been studied. The results show that the Dy utilization efficiency of diffusion method is superior to the addition method, results which are significantly meaningful in their application of the science of Nd–Y–Fe–B magnets.

2. Materials and Methods

2.1. Experimental Procedure

Alloy strips with a nominal composition of $(PrNd)_{20.66}Y_{6.88}B_{0.98}M_{8.5}Fe_{bal}$ (wt.%, M = Al, Cu, Co, named as Y25) were prepared by a strip-casting (SC) technique. The alloy strips were crushed into powders with an average particle size of 2.2 μm by hydrogen decrepitation along with N_2-jet milling. The DyH_x powder with an average size of about 1 μm was prepared by hydrogen absorption fragmentating and N_2-jet milling. Then, 1.6 wt.% DyH_x powder was added into the Y25 matrix powder as a modifier. The Y25 and the mixed powders were pressed and oriented under a magnetic field of 2T in a protective nitrogen atmosphere, followed by iso-static compaction under a pressure of 150 MPa. Subsequently, the sintering process was performed at 1060–1080 °C for 2 h in a vacuum atmosphere, followed by a two-step annealing at 900 °C for 2 h and 500 °C for 2 h to obtain the final magnets. In order to fabricate the diffusion matrix, the sintered Y25 magnet was cut into a cylinder with the size of Φ10 mm × 5 mm, and the diffusion source was prepared by mixing the DyH_x powder and alcohol under the mass ratio of 1:1. The Y25 cylinder was immersed into the diffusion source for 5 s to obtain a uniform DyH_x layer on the surface, then the sample was heat treated at 900 °C for 10 h, followed by annealing at 500 °C for 2 h. The final average Dy content of the diffusion magnet was 0.421 wt.%, obtained by using ICP-OES. For ease of description, the Y25 magnet, the 1.6 wt.% DyH_x addition magnet, and the DyH_x diffusion at 900 °C magnets were named as the original magnet, the DyH_x addition magnet, and the DyH_x diffusion magnet, respectively.

2.2. Characterization and Analysis Methods

The room-temperature and elevated-temperature magnetic properties of the magnets were measured using a magnetic measurement system (NIM-500C). The microstructure observation of the samples was performed in the back scattered mode of scanning electron microscopy (SEM, FEI QUANTA 250, FEI Company, Hillsboro, OR, USA). The contents of elements in the magnets were obtained by using energy-dispersive X-ray spectroscopy (EDS). The elemental concentration mapping was conducted using an electron probe microanalyzer (EPMA, JEOL JXA-8100, Tokyo, Japan). The contents of rare earths in the magnets were examined using a glow discharge emission spectrum (GD-OES, Spectruma Analytik GMBH 750HP, Hof, Germany). The irreversible loss of the magnetic flow was investigated with a Helmholtz coil after exposing the samples at elevated temperatures for 0.5 h in open circuit.

3. Results and Discussion

3.1. Magnetic Properties

Figure 1a demonstrates the demagnetization curves of the original magnet, the DyH_x addition magnet, and the DyH_x diffusion magnet. The variation curves of the corresponding B_r, H_{cj}, and $(BH)_{max}$ of the three type magnets are also shown in Figure 1b. The room temperature magnetic properties of the magnets are listed in Table 1. The results show that the coercivity of the addition magnet is 15.51 kOe, which is a slightly higher than that of the diffusion magnet (15.15 kOe), and the coercivities of the two DyH_x treated magnets are much higher than that of the original magnet (9.45 kOe). Unfortunately, the remanences of the addition magnet and the diffusion magnet, respectively, reduce to 12.59 kG and 12.90 kG compared with the original magnet (13.09 kG). Furthermore, the average Dy content in the diffusion magnet, obtained by using ICP-OES, is 0.421 wt.%, which is much lower than that of the addition magnet. In this work, the utilization efficiency of Dy can be defined as the ratio of ΔH_{cj} to the average Dy content in the magnet. Thus, the coercivity increment of the diffusion magnet is significantly improved to 13.5 kOe/(wt.% Dy), which is much higher than that of the addition magnet (about 3.8 kOe/(wt.% Dy)). The previous studies [29,30] indicate that the remanence deteriorates after Dy doping into the Nd–Fe–B magnet due to the anti-ferromagnetic coupling effect between the heavy rare earth Dy and the transition metal Fe. Thus, the DyH_x diffusion magnet has a higher remanence compared to the DyH_x addition magnet. These changes in coercivity and remanence eventually lead to variety in the maximum energy product $(BH)_{max}$. Although the maximum energy products of the addition and diffusion magnets are decreased compared with the original magnet, the $(BH)_{max}$ of the diffusion magnet is much higher than that of the addition magnet. The $(BH)_{max}$ of the diffusion magnet decreased by only 3.2% compared with the original magnet, which shows significant merit for the Y-containing magnet.

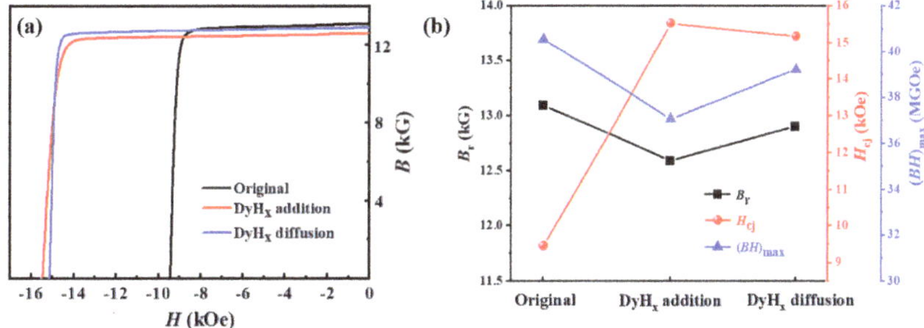

Figure 1. (a) The demagnetization curves of the original magnet, the DyH_x addition magnet, and the DyH_x diffusion magnet; (b) the variation curves of the corresponding B_r, H_{cj}, and $(BH)_{max}$ of the three types of magnets derived from (a).

Table 1. The room temperature magnetic properties of the magnets.

Sample	B_r (kG)	H_{cj} (kOe)	$(BH)_{max}$ (MGOe)
Original	13.09	9.45	40.52
DyH_x addition	12.59	15.51	37.05
DyH_x diffusion	12.90	15.15	39.20

3.2. Microstructure and Elemental Distribution

Figure 2 shows the distribution of Dy element in the DyH_x addition, and diffusion magnets from the surface to an interior of 200 μm depth. The Dy distributes uniformly in the DyH_x addition magnet, and the Dy content is about 1.6 wt.% in the whole magnet. In the DyH_x diffusion magnet, Dy concentration is inhomogeneous and decreases with the increasing diffusion depth. From the surface to the interior about 30 μm of the magnet, Dy concentration dramatically drops with the increasing depth, but it is much higher than that of the addition magnet. At 30 μm depth from the surface of the magnets, the Dy element concentration of the diffusion magnet starts to be lower than that of the addition magnet, and their difference becomes larger with the increasing depth. It can be considered that the utilization efficiency of heavy rare earth Dy by the grain boundary diffusion method is much higher than that of dual alloy method with the same coercivity enhancement.

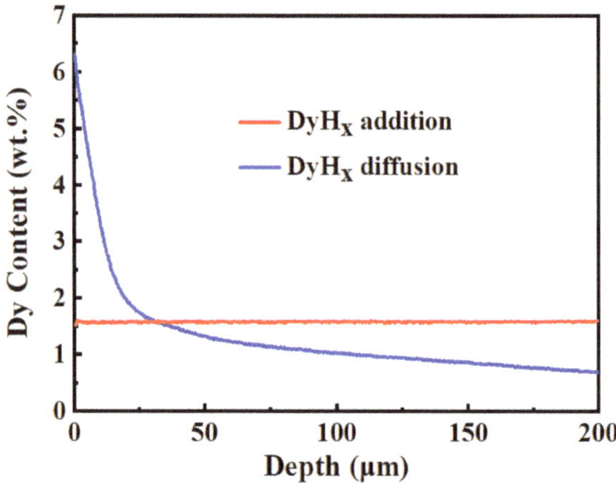

Figure 2. The Dy concentration distributions of the DyH$_x$ addition and diffusion magnets in the depth range of 0–200 μm.

The microstructures of the DyH$_x$ addition magnet and the DyH$_x$ diffusion magnet have also been obtained by using SEM, and are shown in Figure 3. Figure 3(a1,a2) are two different positions of the original magnet, which are shown that the core–shell structure with Y-rich core and Y-lean shell was formed during the heat treatment processes. Figure 3(b1–b3) represent the microstructures of three random positions in the DyH$_x$ addition magnet, which clearly demonstrates that the microstructure inside the addition magnet remains basically the same. Additionally, the Dy-rich shells have been formed in the outer layer of the matrix grains. Figure 3(c1–c4) are the microstructures at the depths of 0 μm, 50 μm, 100 μm, and 200 μm from the surface of the DyH$_x$ diffusion magnet. The results show that the microstructure of the magnet gradually varies from the surface to the interior. The Dy-rich shells are also formed in the outer layer of the matrix grains of the DyH$_x$ diffusion magnet, while the thickness of the Dy-rich shell gradually decreases from the surface to the interior. Furthermore, the thickness of the Dy-rich shell in the DyH$_x$ diffusion magnet is far less than that of the DyH$_x$ addition magnet, which means that the DyH$_x$ grain boundary diffusion method consumes much lower heavy rare earth Dy compared to the dual alloy doping method. On the other hand, the core–shell structure of the DyH$_x$ addition magnet is similar to the original magnet, except that the shell contains the Dy element, while a double shell structure was formed in the DyH$_x$ diffusion magnet that will be discussed below. Moreover, it is also obvious that the grain boundary phase of the DyH$_x$ diffusion magnet is clearer and more continuous than that of the DyH$_x$ addition magnet, which also contributes to the improvement of coercivity.

The distribution of Dy in the magnet plays a significant role in the improvement of magnetic properties and therefore, the detailed elemental distributions of the addition and diffusion magnets was investigated, as shown in Figure 4. The results show that, in the DyH$_x$ addition magnet, large amount Dy atoms diffuse into the matrix grains to form (Nd, Dy)$_2$Fe$_{14}$B shells, which can greatly enhance the surface magnetocrystalline anisotropy field of the matrix grains to improve the coercivity. However, there are also large numbers of Dy atoms agglomerating in the triple junction area, which is unfavorable to the improvement of coercivity. In the DyH$_x$ diffusion magnet, Dy atoms diffuse through grain boundaries into the interior of the magnet. On one hand, Dy atoms diffusion into the surface of the matrix grains to form a thinner (Nd, Dy)$_2$Fe$_{14}$B layer and, thus, fewer heavy rare earth atoms enter the matrix grain to decrease the saturation magnetization of the main phase, leading to the higher remanence of the final diffusion magnet. On the other hand, partial Dy atoms infiltrate into the deeper part inside the magnet along the grain boundary to broaden the thickness of thin grain boundary, resulting in a stronger magnetic isolation effect between the adjacent matrix grains and higher coercivity for the final diffusion magnet.

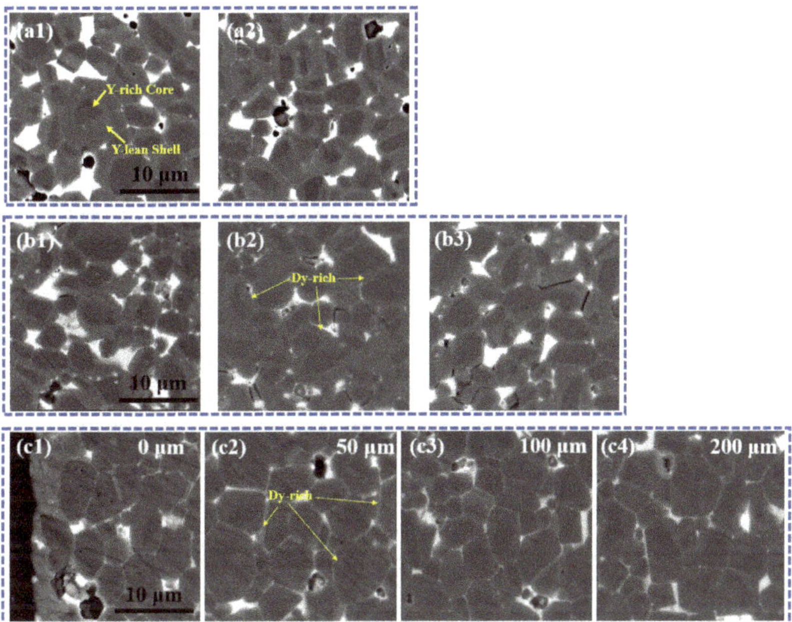

Figure 3. The microstructures of the original magnet, DyH$_x$ addition magnet, and the DyH$_x$ diffusion magnet. (**a1,a2**) are two different positions of the original magnet, (**b1–b3**) are three random positions in the DyH$_x$ addition magnet, and (**c1–c4**) are the microstructures at the depths of 0 μm, 50 μm, 100 μm, and 200 μm from the surface of the DyH$_x$ diffusion magnet, respectively.

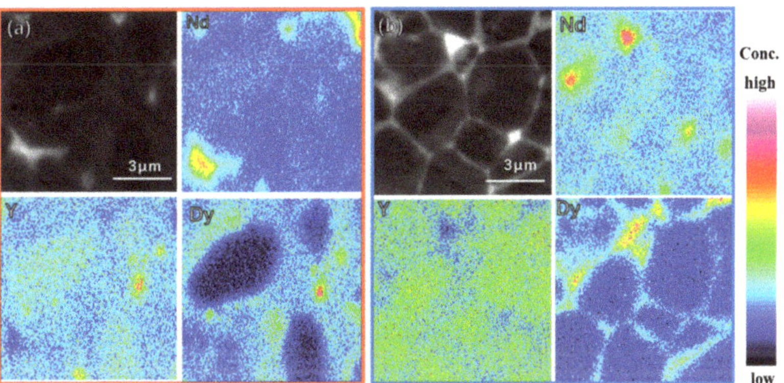

Figure 4. EPMA elemental mappings of the DyH$_x$ addition magnet (**a**) and the DyH$_x$ diffusion magnet (**b**).

Figure 5 shows the elemental distribution of Dy in the matrix grain at a depth of 50 μm for the DyH$_x$ diffusion magnet. It can be seen that three layers with different contrasts appear in the matrix grains after the grain boundary diffusion process. In addition to the core–shell structure where Y forms a Y-rich core and a Y-lean shell in the interior of the matrix grain, Dy also diffuses into the outside of the Y-lean shell to form a Dy-rich shell. The Dy-rich shell is brighter in SEM backscatter image due to the larger atomic weight of Dy than those of Nd and Y. From the center to the surface, the matrix grain forms a double-shell structure with a Y-rich core, a Y-lean Dy-lean shell as well as a Y-lean Dy-rich shell in which the anisotropy field is sequentially enhanced. Therefore, the anisotropy field of the matrix grain is improved after the diffusion process and, thus, the coercivity has been increased.

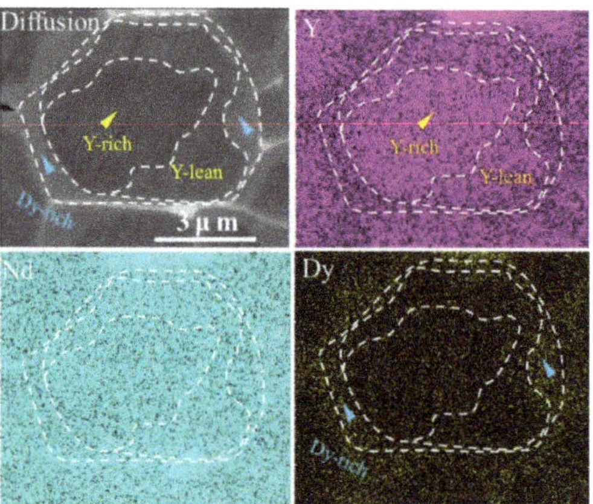

Figure 5. Back-scattered SEM image mappings for the DyH$_x$ diffusion magnet at the depth of 50 μm.

3.3. Thermal Stability

When Nd–Fe–B magnets are used in motors and electronic products, they always need to be able to perform at high temperature due to the heating of the devices, so the magnetic performances of the magnet at high temperatures are very important performance indicators. The temperature stability of Nd–Fe–B magnet is closely related to their microstructures and intrinsic properties. In general, high coercivity is beneficial for magnet temperature stability. Therefore, the dependences of coercivity on the temperature of the original magnet, the DyH$_x$ addition magnet, and the DyH$_x$ diffusion magnet have been obtained, as shown in Figure 6. The temperature coefficients of the coercivity from 20 °C to 120 °C have also been calculated using the formula $\beta = \frac{H_T - H_{T_0}}{H_{T_0}(T - T_0)} * 100\%$, where β is the coercivity temperature coefficient from T_0 to T, which is shown in Table 2.

Figure 6 shows that the coercivities of the magnets decrease with the increasing temperature. While at the same temperature, the coercivities of the DyH$_x$-treated magnets are much higher than that of the original magnet. In the range of 20–120 °C, the coercivity temperature coefficient β of the original magnet is −0.5968%/°C. After introducing DyH$_x$, the coercivity temperature stability has been improved significantly. The β of the DyH$_x$ addition and diffusion magnets are, respectively. −0.5632%/°C and −0.5614%/°C, which means that the coercivity temperature stability of the DyH$_x$ diffusion magnet is slightly better than that of the DyH$_x$ addition magnet.

The irreversible loss of the magnetic flux is also an important parameter to evaluate the high temperature performance of the magnet. As the temperature increases, the magnetic flux will also decrease, but the magnetic flux will recover after the temperature drops to room temperature, thus, the irreversible flux part is called the irreversible magnetic flux loss. Moreover, the smaller the irreversible magnetic flux loss of the magnet, the better the high temperature resistance of the magnet. Figure 7 shows the irreversible flux loss versus aging temperature for the initial magnet, the DyH_x addition magnet, and the DyH_x diffusion magnet. The results show that the irreversible flux loss of the initial magnet is about 20% after being treated at 80 °C for 2 h. However, the irreversible flux losses of the DyH_x-treated magnets significantly reduce, which means that the loss values of DyH_x addition and diffusion magnet are 2.6% and 0.43%, respectively. Moreover, the irreversible flux loss of the DyH_x addition magnet is about 10% when the temperature increases to 100 °C, but the same loss value has been obtained in the DyH_x diffusion magnet at higher that 120 °C, indicating the enhanced high temperature stability of the DyH_x diffusion treatment compared to DyH_x addition.

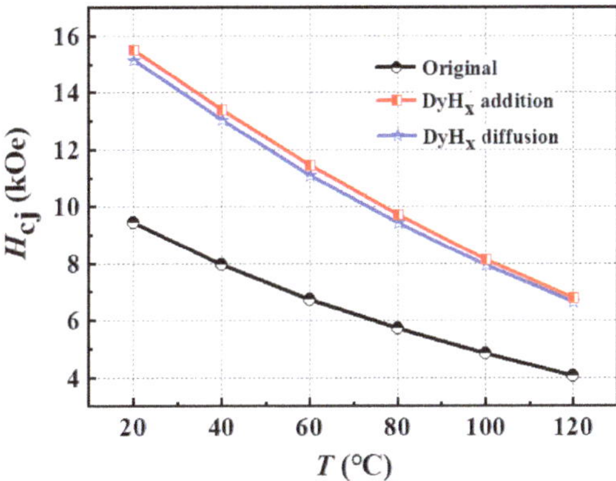

Figure 6. The coercivity versus temperature of the original magnet, the DyH_x addition magnet, and the DyH_x diffusion magnet.

Table 2. The coercivity temperature coefficient β from 20 °C to 120 °C of the magnets.

Magnets	β (%/°C)
Original	−0.5968
DyH_x addition	−0.5632
DyH_x diffusion	−0.5614

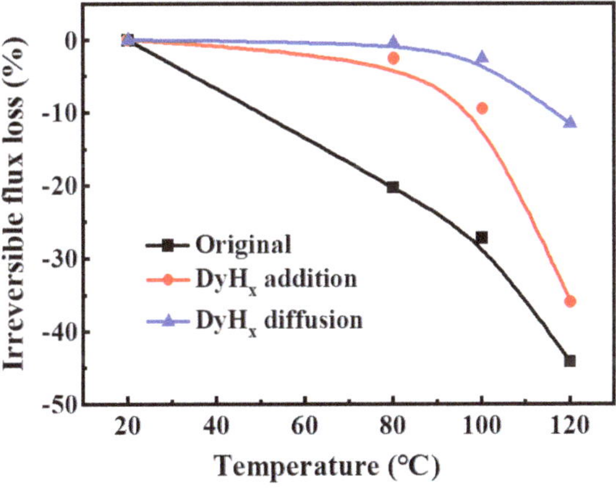

Figure 7. Irreversible loss of flux versus temperature of the original magnet, the DyH_x addition magnet, and the DyH_x diffusion magnet.

3.4. Discussions

The above investigations show that the magnetic performances and the high temperature stability of the DyH_x diffusion magnet is higher than that of the DyH_x addition magnet, which is mainly attributed to the differences of microstructures and elemental distributions due to the way that Dy enters the magnets. When they are used as a grain boundary additive, the Dy atoms generate liquid phases to fill the gaps between grains and aggregate in the matrix phase grain intersection areas to form triple junction phases during the sintering process. While they are treated as diffusion sources, the Dy atoms diffuse through the melted grain boundary phase during the diffusion process and do not accumulate at the grain boundary. At the same time, the melting Dy repairs the defects among the matrix grain and the grain boundary phase, and the distance between the matrix grain is also broadened.

The Dy-rich shells which are formed in the matrix grains of DyH_x addition and diffusion magnets are also significantly different. The thickness of the $(Nd, Dy)_2Fe_{14}B$ shell in the addition magnet is higher than that of diffusion magnet, which is mainly due to the fact that the Dy atom at the grain boundary is more likely to enter the matrix grains and be substituted by an Nd atom at elevated temperatures. In the grain boundary diffusion process, a thin Dy-rich shell is formed in the surface of the matrix grain, which greatly improves the utilization of the Dy element. Moreover, it also prevents too much Dy from entering the matrix grains to reduce the saturation magnetization. Therefore, the grain boundary diffusion method of Dy has a higher utilization efficiency in improving the coercivity.

4. Conclusions

We investigated the effect of DyH_x addition and diffusion methods on the microstructure, magnetic performance, and thermal stability of the Nd–Y–Fe–B magnet. The coercivity of the original magnet increased from 9.45 kOe to 15.51 kOe for the DyH_x addition magnet and 15.15 kOe for the DyH_x diffusion magnet. However, the coercivity increment of the Dy element of the diffusion method was up to 13.5 kOe/(wt.% Dy), much higher than the addition method (about 3.8 kOe/(wt.% Dy)). Moreover, the remanence of the DyH_x diffusion magnet was as high att 12.90 kG, which was better than the DyH_x addition magnet (12.59 kG). These superior magnetic performances of DyH_x diffusion magnet are mainly due to the outstanding utilization efficiency of Dy that diffused into the outer layer of the matrix grain to form a thinner Dy-rich shell and infiltrated along the grain boundary to construct a clear and continuous grain boundary phase. The Dy-rich shell in the DyH_x addition magnet was similar, with the original core–shell structure in the Nd–Y–Fe–B magnet. However, the distinct dual-shell structure consisting of a thinner Dy-rich shell and Y-lean shell was constructed in the DyH_x diffused magnet, contributing to the superior coercivity increment and Dy utilization efficiency. Based on the distribution characteristic of Dy in the magnet, the thermal stability of the DyH_x diffusion magnet is also superior to the DyH_x addition magnet, which will greatly expand the application range of the Nd–Y–Fe–B magnet.

Author Contributions: Investigation, formal analysis, writing and editing, S.G. and X.F.; Materials preparation and testing, S.L. and G.D.; Materials preparation and writing—review and editing, B.Z.; project administration, funding acquisition, R.C. and A.Y.; All authors have read and agreed to the published version of the manuscript.

Funding: This research was funded by Key R&D program of Zhejiang Province (No. 2021C01190), Major Project of "Science and Technology Innovation 2025" in Ningbo City (No. 2020Z046), Inner Mongolia Major Technology Project (2019ZD020), Kunpeng plan of Zhejiang Province and Ningbo top talent program.

Institutional Review Board Statement: Not applicable.

Informed Consent Statement: Not applicable.

Data Availability Statement: The data presented in this study are available on request from the corresponding author.

Conflicts of Interest: The authors declare no conflict of interest.

References

1. Jones, N. The pull of stronger magnets. *Nature* **2011**, *472*, 22. [CrossRef] [PubMed]
2. Sagawa, M.; Fujimura, S.; Togawa, N.; Yamamoto, H.; Matsuura, Y. New material for permanent magnets on a base of Nd and Fe. *J. Appl. Phys.* **1984**, *55*, 2083. [CrossRef]
3. Gutfleisch, O.; Willard, M.A.; Brück, E.; Chen, C.H.; Sankar, S.G.; Liu, J.P. Magnetic materials and devices for the 21st century: Stronger, lighter, and more energy efficient. *Adv. Mater.* **2011**, *23*, 821. [CrossRef]
4. Herbst, J.F. $Re_2Fe_{14}B$ materials: Intrinsic properties and technological aspects. *Rev. Mod. Phys.* **1991**, *63*, 819. [CrossRef]
5. Ding, G.F.; Guo, S.; Chen, L.; Ding, J.H.; Song, J.; Chen, R.J.; Lee, D.; Yan, A. Coercivity enhancement in Dy-free sintered Nd-Fe-B magnets by effective structure optimization of grain boundaries. *J. Alloys Compd.* **2018**, *735*, 795. [CrossRef]
6. Coey, J.M.D. Hard magnetic materials: A perspective. *IEEE Trans. Magn.* **2011**, *47*, 4671. [CrossRef]
7. Cao, X.J.; Chen, L.; Li, X.B.; Yi, P.P.; Yan, A.R.; Yan, G.L. Coercivity enhancement of sintered Nd-Fe-B magnets by efficiently diffusing DyF_3 based on electrophoretic deposition. *J. Alloys Compd.* **2015**, *631*, 315. [CrossRef]
8. Ding, G.F.; Guo, S.; Chen, L.; Di, J.H.; Chen, K.; Chen, R.J.; Lee, D.; Yan, A. Effects of the grain size on domain structure and thermal stability of sintered Nd-Fe-B magnets. *J. Alloys Compd.* **2018**, *735*, 1176–1180. [CrossRef]
9. Kim, T.H.; Lee, S.R.; Kim, H.J.; Lee, M.W.; Jiang, T.S. Simultaneous application of Dy-*X* (*X* = F or H) powder doping and dip-coating processes to Nd-Fe-B sintered magnets. *Acta Mater.* **2015**, *93*, 95–104. [CrossRef]
10. Li, W.F.; Sepehri-Amin, H.; Ohkubo, T.; Hase, N.; Hono, K. Distribution of Dy in high-coercivity (Nd, Dy)-Fe-B sintered magnet. *Acta Mater.* **2011**, *59*, 3061. [CrossRef]
11. Zhang, M.; Li, Z.B.; Shen, B.G.; Hu, F.X.; Sun, J.R. Permanent magnetic properties of rapidly quenched $(La,Ce)_2Fe_{14}B$ nanomaterials based on La-Ce mischmetal. *J. Alloys Comp.* **2015**, *651*, 144. [CrossRef]
12. Pathak, A.K.; Khan, M.; Gschneidner, J.K.A.; McCallum, R.W.; Zhou, L.; Sun, K.; Dennis, K.W.; Zhou, C.; Pinkerton, F.E.; Kramer, M.J.; et al. Cerium: An unlikely replacement of Dysprosium in high performance Nd-Fe-B permanent magnets. *Adv. Mater.* **2015**, *27*, 2663. [CrossRef] [PubMed]
13. Yan, C.J.; Guo, S.; Chen, R.J.; Lee, D.; Yan, A.R. Enhanced magnetic properties of sintered Ce-Fe-B based magnets by optimizing the microstructure of strip-casting alloys. *IEEE Trans. Magn.* **2014**, *50*, 2104604. [CrossRef]
14. Ma, T.Y.; Wu, B.; Zhang, Y.J.; Jin, J.Y.; Wu, K.Y.; Tao, S.; Xia, W.; Yan, M. Enhanced coercivity of NdCe-Fe-B sintered magnets by adding (Nd, Pr)-H powders. *J. Alloys Compd.* **2017**, *721*, 1. [CrossRef]
15. Fan, X.D.; Ding, G.F.; Chen, K.; Guo, S.; You, C.Y.; Chen, R.J.; Lee, D.; Yan, A. Whole process metallurgical behavior of the high-abundance rare-earth elements LRE (La, Ce and Y) and the magnetic performance of $Nd_{0.75}LRE_{0.25}$-Fe-B sintered magnets. *Acta Mater.* **2018**, *154*, 343–354. [CrossRef]
16. Ding, G.F.; Liao, S.C.; Di, J.H.; Zheng, B.; Guo, S.; Chen, R.J.; Yan, A. Microstructure of core-shell NdY-Fe-B sintered magnets with a high coercivity and excellent thermal stability. *Acta Mater.* **2020**, *194*, 547–557. [CrossRef]
17. Fan, X.D.; Chen, K.; Guo, S.; Chen, R.J.; Lee, D.; Yan, A.; You, C.Y. Core–shell Y-substituted Nd–Ce–Fe–B sintered magnets with enhanced coercivity and good thermal stability. *Appl. Phys. Lett.* **2017**, *110*, 172405. [CrossRef]
18. Yang, F.; Guo, L.C.; Li, P.; Zhao, X.Z.; Sui, Y.L.; Guo, Z.M.; Gao, X.X. Boundary structure modification and magnetic properties of Nd-Fe-B sintered magnets by co-doping with Dy_2O_3/S powders. *J. Magn. Magn. Mater.* **2017**, *429*, 117–123. [CrossRef]
19. Cui, X.G.; Cui, C.Y.; Cheng, X.N.; Xu, X.J. Effect of Dy_2O_3 intergranular addition on thermal stability and corrosion resistance of Nd-Fe-B magnets. *Intermetallics* **2014**, *55*, 118–122. [CrossRef]
20. Zhao, Y.; Feng, H.B.; Li, A.H.; Li, W. Microstructural and magnetic property evolutions with diffusion time in TbH_x diffusion processed Nd-Fe-B sintered magnets. *J. Magn. Magn. Mater.* **2020**, *515*, 167272. [CrossRef]
21. Liu, P.; Ma, T.Y.; Wang, X.H.; Zhang, Y.J.; Yan, M. Role of hydrogen in Nd–Fe–B sintered magnets with DyH_x addition. *J. Alloys Compd.* **2015**, *628*, 282–286. [CrossRef]
22. Wang, C.G.; Yue, M.; Zhang, D.T.; Liu, W.Q.; Zhang, J.X. Structure and magnetic properties of hot deformed $Nd_2Fe_{14}B$ magnets doped with DyH_x nanoparticles. *J. Magn. Magn. Mater.* **2016**, *404*, 64–67. [CrossRef]
23. Bae, K.-H.; Kim, T.-H.; Lee, S.-R.; Kim, H.-J.; Lee, M.-W.; Jang, T.-S. Magnetic and microstructural characteristics of DyF_3/DyH_x dip-coated Nd–Fe–B sintered magnets. *J. Alloys Compd.* **2014**, *612*, 183–188. [CrossRef]
24. Wang, C.; Luo, Y.; Wang, Z.L.; Yan, W.L.; Zhao, Y.Y.; Quan, N.T.; Peng, H.J.; Wu, K.W.; Ma, Y.H.; Zhao, C.L.; et al. Effect of $MgCl_2$ on electrophoretic deposition of TbF_3 powders on Nd-Fe-B sintered magnet. *J. Rare Earths* **2022**, in press. [CrossRef]
25. Oono, N.; Sagawa, M.; Kasada, R.; Matsui, H.; Kimura, A. Production of thick high-performance sintered neodymium magnets by grain boundary diffusion treatment with dysprosium–nickel–aluminum alloy. *J. Magn. Magn. Mater.* **2011**, *323*, 297–300. [CrossRef]
26. Xu, F.; Wang, J.; Dong, X.P.; Zhang, L.T.; Wu, J.S. Grain boundary microstructure in DyF_3 -diffusion processed Nd–Fe–B sintered magnets. *J. Alloys Compd.* **2011**, *509*, 7909–7914. [CrossRef]

27. Lv, M.; Kong, T.; Zhang, W.H.; Zhu, M.Y.; Jin, H.M.; Li, W.X.; Li, Y. Progress on modification of microstructures and magnetic properties of Nd- Fe-B magnets by the grain boundary diffusion engineering. *J. Magn. Magn. Mater.* **2021**, *517*, 167278. [CrossRef]
28. Soderžnik, M.; Korent, M.; Soderžnik, K.Ž.; Katter, M.; Üstüner, K.; Kobe, S. High-coercivity Nd-Fe-B magnets obtained with the electrophoretic deposition of sub-micron TbF_3 followed by the grain-boundary diffusion process. *Acta Mater.* **2016**, *115*, 278–284. [CrossRef]
29. Löewe, K.; Brombacher, C.; Katter, M.; Gutfleisch, O. Temperature-dependent Dy diffusion processes in Nd–Fe–B permanent magnets. *Acta Mater.* **2015**, *83*, 248–255. [CrossRef]
30. Sepehri-Amin, H.; Ohkubo, T.; Hono, K. The mechanism of coercivity enhancement by the grain boundary diffusion process of Nd–Fe–B sintered magnets. *Acta Mater.* **2013**, *61*, 1982–1990. [CrossRef]

 *materials*

Article

Transport Property and Spin–Orbit Torque in 2D Rashba Ferromagnetic Electron Gas

Chao Yang [1,*], Da-Kun Zhou [2], Ya-Ru Wang [2] and Zheng-Chuan Wang [2,*]

[1] College of Mechanical and Electrical Engineering, Wuyi University, Wuyishan 354300, China
[2] School of Physical Sciences, University of Chinese Academy of Sciences, Beijng 100049, China; zhoudakun17@mails.ucas.edu.cn (D.-K.Z.); wangyaru18@mails.ucas.edu.cn (Y.-R.W.)
[*] Correspondence: wyxy_cyang@163.com (C.Y.); wangzc@ucas.ac.cn (Z.-C.W.)

Abstract: In this paper, we investigate the spin–orbit torque and transport property in a 2D Rashba ferromagnetic electron gas. The longitudinal conductivity can be divided into two parts: the first term is determined by the charge density and is independent of the spin degrees of freedom. The second term depends on the two bands that spin in the opposite directions, and it is directly proportional to spin–orbit torque regardless of the band structure and temperature. This is a general and underlying relation between the transport property and spin–orbit torque. Moreover, we show the impacts of the spin–orbit coupling constant and Fermi energy on transverse conductivity and spin–orbit torque, which is helpful for relevant experiments.

Keywords: 2D Rashba ferromagnetic electron gas; spin-orbit torque; longitudinal conductivity

PACS: 72.25.-b; 75.25.-j; 75.76.+j; 85.75.-d

Citation: Yang, C.; Zhou, D.-K.; Wang, Y.-R.; Wang, Z.-C. Transport Property and Spin–Orbit Torque in 2D Rashba Ferromagnetic Electron Gas. *Materials* **2022**, *15*, 5149. https://doi.org/10.3390/ma15155149

Academic Editor: George Kioseoglou

Received: 8 June 2022
Accepted: 19 July 2022
Published: 25 July 2022

Publisher's Note: MDPI stays neutral with regard to jurisdictional claims in published maps and institutional affiliations.

1. Introduction

In recent years, spin–orbit torque has become one of the most attractive topics in the field of spintronics because of its great application prospects in magnetic information storage technology [1–3]. spin–orbit torque (SOT) is based on spin–orbit interaction (SOI), which uses the non-equilibrium spin accumulation induced by charge flow to generate a torque on a local magnetic moment [4,5], thus achieving the purpose of regulating magnetic storage units. SOT-based magnetic random access memory (SOT-MRAM) overcomes the disadvantages of STT-MRAM, especially the fact that it separates the read and write paths, so it has a higher read and write speed and lower power consumption than STT-MRAM [6,7]. Relevant studies show that SOT-MRAM can achieve ultra-fast information writing, which is reduced from the tens of nanoseconds required by the original STT-MRAM to less than 10 ns, while the power consumption of the device is further reduced [8,9].

The essence of spin–orbit torque is that the directional movement of electrons produces a non-equilibrium spin accumulation in the spin–orbit coupling system, by means of the s–d interaction, the spin accumulation of conducting electrons applies a torque on the local magnetic moment [10,11]. Meanwhile, this directional movement of electrons also generates a charge current in general. Thus, the current and spin–orbit torque are both manifestations of the non-equilibrium transport properties of the SOC system, there must be some relations between them. Literature have showed that the magnitude of spin–orbit torque is often proportional to the current density [12–15]. Because the spin accumulation and charge current are proportional to the external electric field under linear transport conditions, the ratio depends on the band structure of the system [4,5]. There is no clear conclusion about the underlying relationship between spin–orbit torque and conductivity.

Two-dimensional Rashba ferromagnetic electron gas is an ideal platform for investigating spin–orbit torque [16,17], and the latter is often realized at heavy metal/ferromagnetic interfaces [18,19]. These systems have both Rashba spin–orbit interaction and ferromagnetism, and interesting transport phenomena such as spin–orbit torque and anomalous Hall effect have been found [20,21]. In this paper, we use 2D Rashba ferromagnetic electron gas as an example to explore the relationship between spin–orbit torque and transport properties, as well as the regulation of spin–orbit torque. In Section 2, we show the energy splitting caused by spin–orbital interactions; in Section 3, we study the transport properties of the system, including longitudinal conductance and intrinsic anomalous Hall conductivity; in Section 4, we investigate the spin–orbit torque, and show its regulation in terms of the spin–orbit coupling constant and Fermi energy; Section 5 is the conclusion.

2. Model

The Hamiltonian of an electron in a 2D Rashba ferromagnetic electron gas is $\hat{H} = \frac{\hbar^2 \vec{k}^2}{2\mu} + \alpha(\vec{k} \times \vec{z}) \cdot \hat{\vec{\sigma}} - J\vec{M} \cdot \hat{\vec{\sigma}}$, where μ is the effective mass of the electron, $\vec{M}$ is the direction of magnetization and we take an out-of-plane magnetization as: $\vec{M} = (0,0,1)$, J and α are the constant of the s–d interaction and Rashba spin–orbit coupling. To be sure, the s–d interaction is generated between the electron gas (s electrons) and the local magnetic moments (d electrons), where the latter form a three-dimensional ferromagnetic layer [4,5]. Therefore, the Hamiltonian can be written as [16,17,21]:

$$\hat{H} = \begin{pmatrix} \frac{\hbar^2 k^2}{2\mu} - J & \alpha k_y + i\alpha k_x \\ \alpha k_y - i\alpha k_x & \frac{\hbar^2 k^2}{2\mu} + J \end{pmatrix} \tag{1}$$

Solving the Schrodinger's equation, we obtain

$$\epsilon_{\uparrow,\downarrow}(\vec{k}) = \frac{\hbar^2 k^2}{2\mu} \pm \sqrt{J^2 + \alpha^2 k^2}, \tag{2}$$

where $\Delta = \sqrt{J^2 + \alpha^2 k^2}$ is the energy splitting of these two bands. Correspondingly, the wavefunctions are $\psi_{\uparrow,\downarrow}(\vec{k}) = e^{i\vec{k}\cdot\vec{r}} | \uparrow, \downarrow\rangle_{\vec{k}}$, where

$$| \uparrow, \downarrow\rangle_{\vec{k}} = \frac{1}{\sqrt{(J \pm \Delta)^2 + \alpha^2 k^2}} \begin{pmatrix} \alpha k_y + i\alpha k_x \\ J \pm \Delta \end{pmatrix}. \tag{3}$$

Then, the average spin of the state $| \uparrow\rangle_{\vec{k}}$ is $\vec{s}(\vec{k}) =_{\vec{k}} \langle \uparrow |\hat{\vec{\sigma}}| \uparrow\rangle_{\vec{k}} = (\frac{\alpha k_y}{\Delta}, \frac{-\alpha k_x}{\Delta}, \frac{J}{\Delta})$, and the average spin of the state $| \downarrow\rangle_k$ is $_{\vec{k}}\langle \downarrow |\hat{\vec{\sigma}}| \downarrow\rangle_{\vec{k}} = -\vec{s}(\vec{k})$.

Figure 1 shows the bands splitting caused by spin–orbit interaction. The spins of electrons in these two bands are in opposite directions, and the energy difference is 2Δ. The band structure depends on the relative magnitudes of α and J. If $\alpha k_J < J$, where $k_J = \sqrt{2\mu J}/\hbar$, $\epsilon_\uparrow$ and $\epsilon_\downarrow$ are both parabolic. If $\alpha k_J > J$, $\epsilon_\downarrow$ has a maximum at $k = 0$ and two bottoms. The specific band structures will affect transport property.

(a)

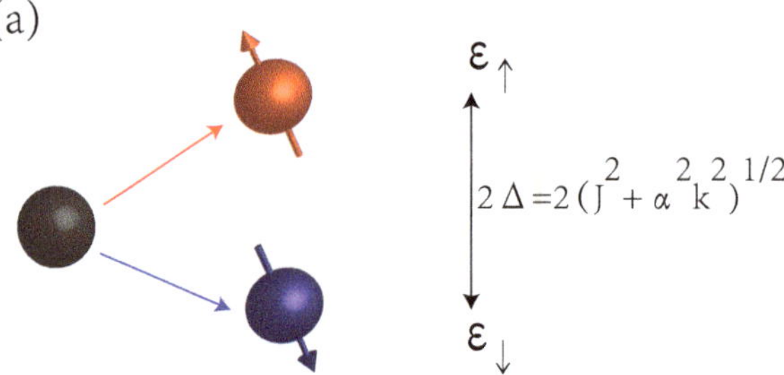

Figure 1. *Cont.*

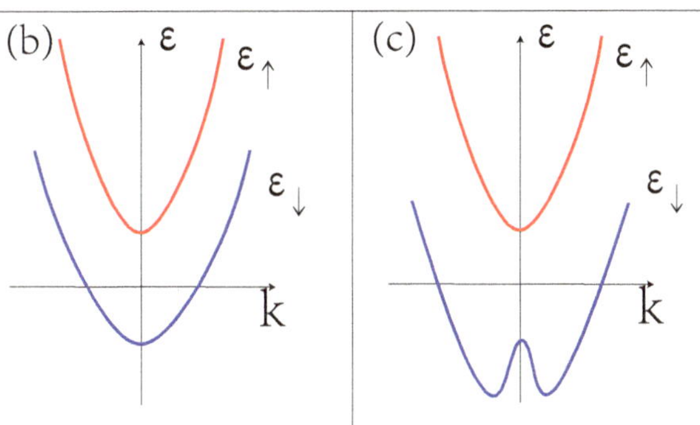

Figure 1. (**a**) Bands splitting with opposite spins; and bands diagram when (**b**) $\alpha k_J < J$, (**c**) $\alpha k_J > J$.

3. Transport Property

In a semiclassical transport equation, the velocity of electrons in band n can be expressed as $v_n = \frac{1}{\hbar}\frac{\partial \epsilon_{kn}}{\partial k} - \frac{e\vec{E}}{\hbar} \times \vec{\Omega}_{kn}$, where $\frac{1}{\hbar}\frac{\partial \epsilon_{kn}}{\partial k}$ is the traditional velocity and $-\frac{e\vec{E}}{\hbar} \times \vec{\Omega}_{kn}$ is the anomalous velocity arising from Berry curvature [22]. In our system, the band splits into $\epsilon_{\uparrow,\downarrow}$, therefore the velocities of electrons are:

$$\vec{v}_{\uparrow,\downarrow} = \frac{1}{\hbar}\vec{\nabla}_k \epsilon_{\uparrow,\downarrow} - \frac{e\vec{E}}{\hbar} \times \vec{\Omega}_{\uparrow,\downarrow}, \tag{4}$$

where $\vec{E}$ is the external electric field and the Berry curvature [22] $\vec{\Omega}_{\uparrow,\downarrow}$ can be calculated from wavefunctions: $\vec{\Omega}_{\uparrow,\downarrow}(\vec{k}) = \pm\frac{J\alpha^2}{2\Delta^3}\vec{z}$.

In a weak external electric field, the distribution functions are $f_{\uparrow,\downarrow} = f^0_{\uparrow,\downarrow} + f^1_{\uparrow,\downarrow}$, where $f^0_{\uparrow,\downarrow}$ are the equilibrium distribution functions, and $f^1_{\uparrow,\downarrow}$ is the first-order perturbation caused by an external electric field. By the relaxation time approximation, $f^1_{\uparrow,\downarrow}$ can be expressed as: $\frac{e\tau}{\hbar}\vec{E} \cdot \vec{\nabla}_k f^0_{\uparrow,\downarrow}$ and τ is the momentum relaxation time [4]. By the definition of current density $\vec{j} = e\int(\vec{v}_\uparrow f_\uparrow + \vec{v}_\downarrow f_\downarrow)d\vec{k}$, longitudinal and transverse current densities are:

$$j_x = \frac{e\hbar}{\mu}\int k_x(f^1_\uparrow + f^1_\downarrow)d\vec{k} + \frac{e\alpha^2}{\hbar}\int \frac{k_x}{\Delta}(f^1_\uparrow - f^1_\downarrow)d\vec{k}, \tag{5}$$

and

$$j_y = \frac{J\alpha^2 e^2 E}{2\hbar}\int \frac{1}{\Delta^3}(f^0_\uparrow - f^0_\downarrow)d\vec{k}. \tag{6}$$

To simplify matters, the equilibrium distribution functions can be expressed via the step function at 0K: $f^0_{\uparrow,\downarrow} = \theta(\epsilon_F - \epsilon_{\uparrow,\downarrow})$, where ϵ_F is the Fermi energy. Thus, the longitudinal conductivity is

$$\sigma_{xx} = \frac{e^2\tau}{\mu}\int k_x^2\left(\frac{\partial f^0_\uparrow}{\partial k_x} + \frac{\partial f^0_\downarrow}{\partial k_x}\right) + \frac{e^2\alpha^2\tau}{\hbar^2}\int \frac{k_x^2}{\Delta}\left(\frac{\partial f^0_\uparrow}{\partial k_x} - \frac{\partial f^0_\downarrow}{\partial k_x}\right), \tag{7}$$

where the first term of Equation (7) can be calculated as:

$$\sigma^1_{xx} = \frac{e^2\tau}{\mu}\int k_x^2\left(\frac{\partial f^0_\uparrow}{\partial k_x} + \frac{\partial f^0_\downarrow}{\partial k_x}\right)d\vec{k}$$

$$= \frac{\pi e^2\tau}{\mu}(k_{F\uparrow}^2 + k_{F\downarrow}^2), \tag{8}$$

where $k_{F\uparrow,\downarrow}$ is calculated with the formulas: $\frac{\hbar^2 k^2}{2\mu} \pm \sqrt{J^2 + \alpha^2 k^2} = \epsilon_F$. Allowing for the charge density $n = \int (f_\uparrow^0 + f_\downarrow^0) = \pi(k_{F\uparrow}^2 + k_{F\downarrow}^2)$, σ_{xx}^1 can be expressed as $\frac{ne^2\tau}{\mu}$ which has the same form with the conductivity of free electron gas. This conductivity is determined by the charge density of electrons, and independent of the spin degree of freedom.

The second term of Equation (7) is derived from the different velocities in two bands:

$$\sigma_{xx}^2 = \frac{e^2 \alpha^2 \tau}{\hbar^2} \int \frac{k_x^2}{\Delta} \left(\frac{\partial f_\uparrow^0}{\partial k_x} - \frac{\partial f_\downarrow^0}{\partial k_x} \right) d\vec{k}. \tag{9}$$

This conductivity results from the different velocities of spin-up and spin-down electrons, and reveals the spin-polarized transport property. Considering the change in Fermi surface, we discuss the results in categories:

When $\epsilon_F > J$, the results of equations $\frac{\hbar^2 k^2}{2\mu} \pm \sqrt{J^2 + \alpha^2 k^2} = \epsilon_F$ are $k_{F\uparrow}$ and $k_{F\downarrow}$, respectively. Thus,

$$\sigma_{xx}^2 = \frac{\pi e^2 \alpha^2 \tau}{\hbar^2} \left(\frac{k_{F\downarrow}^2}{\sqrt{J^2 + \alpha^2 k_{F\downarrow}^2}} - \frac{k_{F\uparrow}^2}{\sqrt{J^2 + \alpha^2 k_{F\uparrow}^2}} \right). \tag{10}$$

When $-J < \epsilon_F < J$, the result of equation $\frac{\hbar^2 k^2}{2\mu} - \sqrt{J^2 + \alpha^2 k^2} = \epsilon_F$ is $k_{F\downarrow}$. Thus,

$$\sigma_{xx}^2 = \frac{\pi e^2 \alpha^2 \tau}{\hbar^2} \frac{k_{F\downarrow}^2}{\sqrt{J^2 + \alpha^2 k_{F\downarrow}^2}}. \tag{11}$$

When $\epsilon_F < -J$, only if $\alpha k_J > J$, the equation $\frac{\hbar^2 k^2}{2\mu} - \sqrt{J^2 + \alpha^2 k^2} = \epsilon_F$ has the solutions $k_{F\downarrow 1}$ and $k_{F\downarrow 2}$ ($k_{F\downarrow 1} > k_{F\downarrow 2}$). Thus,

$$\sigma_{xx}^2 = \frac{\pi e^2 \alpha^2 \tau}{\hbar^2} \left(\frac{k_{F\downarrow 1}^2}{\sqrt{J^2 + \alpha^2 k_{F\downarrow 1}^2}} - \frac{k_{F\uparrow 2}^2}{\sqrt{J^2 + \alpha^2 k_{F\uparrow 2}^2}} \right). \tag{12}$$

Figure 2 shows the longitudinal conductivity as a function of Fermi energy ϵ_F. With the increase in ϵ_F, the charge density n increases gradually. Thus, σ_{xx}^1 increases monotonously. At $\epsilon_F = \pm 1$, the number of crossover points of the Fermi surface changes. When $\epsilon_F = -1$, the Fermi surface intersects with the top of $\epsilon_\downarrow$ at $k = 0$. Below this point, there are two Fermi wave vectors $k_{F\downarrow 1}$ and $k_{F\downarrow 2}$; above that point, σ_{xx}^2 increases with the only Fermi wave vector $k_{F\downarrow}$. Furthermore, when $\epsilon_F = 1$, the Fermi-surface intersects with the bottom of $\epsilon_\uparrow$. Above this point, electrons in band $\epsilon_\uparrow$ participate in the conduction process. From Equation (7), we can see a negative contribution made by $\epsilon_\uparrow$. Thus, σ_{xx}^2 decreases gradually.

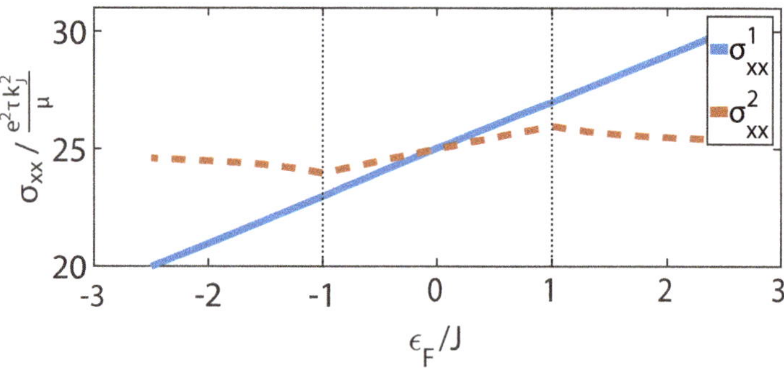

Figure 2. Longitudinal conductivities σ_{xx}^1 and σ_{xx}^2 vs. the Fermi energy ϵ_F, where $\alpha k_J / J = 5$, the solid line is σ_{xx}^1, and the dashed line is σ_{xx}^2.

The transverse conductivity is just the intrinsic anomalous Hall conductance, which originates from the Berry curvature [17]. Accordingly, the transverse conductance is

$$
\sigma_{xy} = \begin{cases} \dfrac{\pi J e^2}{2\hbar}\left(\dfrac{1}{\sqrt{J^2+\alpha^2 k_{F\downarrow}^2}} - \dfrac{1}{\sqrt{J^2+\alpha^2 k_{F\uparrow}^2}} \right) & \epsilon_F > J \\[4mm] \dfrac{\pi J e^2}{2\hbar}\dfrac{1}{\sqrt{J^2+\alpha^2 k_{F\downarrow}^2}} & -J < \epsilon_F < J \\[4mm] \dfrac{\pi J e^2}{2\hbar}\left(\dfrac{1}{\sqrt{J^2+\alpha^2 k_{F\downarrow 1}^2}} - \dfrac{1}{\sqrt{J^2+\alpha^2 k_{F\uparrow 2}^2}} \right) & \epsilon_F > J \end{cases} \tag{13}
$$

Figure 3 shows the transverse conductivity as a function of Fermi energy ϵ_F for different band structures. According to Equation (6), σ_{xy} depends on the equilibrium distribution functions $f^0_{\uparrow,\downarrow}$, which means that all states below Fermi energy contribute to transverse conductivity. For $\alpha k_J < J$, $\epsilon_\uparrow$ and $\epsilon_\downarrow$ are both parabolic. When $\epsilon_F > J$, $\epsilon_\uparrow$ begins to make a negative contribution, and thus σ_{xy} decreases. For $\alpha k_J > J$, $\epsilon_\downarrow$ has a top at $k = 0$ and two bottoms. The maximum of Berry curvature is just at $k = 0$, which means a maximum of transverse velocity. Therefore, the transverse conductivity mainly depends on the states around $k = 0$. When $\epsilon_F < -J$, σ_{xy} increases with the increase in ϵ_F, because the states in $\epsilon_\downarrow$ around $k = 0$ gradually contribute to σ_{xy}. When $-J < \epsilon_F < J$, σ_{xy} is almost unchanged. When $\epsilon_F > J$, σ_{xy} decreases because of the negative contribution by states in $\epsilon_\uparrow$.

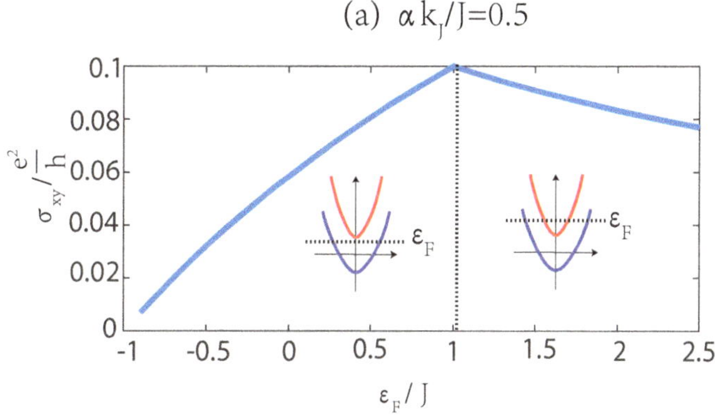

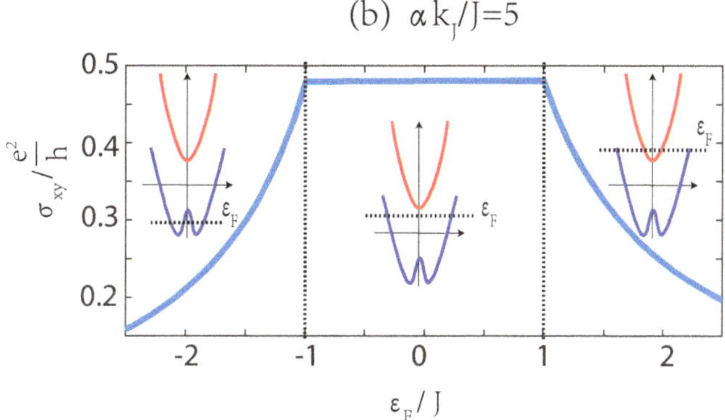

Figure 3. Transverse conductivity σ_{xy} vs. the Fermi energy ϵ_F, where (**a**) $\alpha k_J/J = 0.5$ and (**b**) $\alpha k_J/J = 5$, the red line and blue line are bandstructures of $\epsilon_\uparrow$ and $\epsilon_\downarrow$ respectively.

4. Spin–Orbit Torque

The spin–orbit torque is the result of s–d interaction between the local magnetic moment and spin accumulation of conduction electrons, and it is always expressed as [23,24]:

$$\vec{T} = -\frac{J}{\hbar}\vec{M} \times \vec{m},$$

(14)

where $\vec{m}$ is the spin accumulation and can be calculated by definition [4,15]:

$$\vec{m} = \hbar \int [\vec{s}(\vec{k})f_\uparrow - \vec{s}(\vec{k})f_\downarrow]d\vec{k}.$$

(15)

According to the distribution function above, the spin accumulation along the y axis is

$$m_y = eE\tau\alpha \int \frac{k_x^2}{\Delta}\left(\frac{\partial f_\uparrow^0}{\partial k_x} - \frac{\partial f_\downarrow^0}{\partial k_x}\right)d\vec{k}.$$

(16)

Compared to Equation (9), m_y is directly proportional to σ_{xx}^2. $m_y/\sigma_{xx}^2 = \hbar^2 E/e\alpha$, and this specific value is a constant which has no connection with the band structure and temperature. Thus, the spin accumulation m_y and σ_{xx}^2 have the same variation trend in terms of ϵ_F.

Figure 4 shows the spin accumulation m_y as a function of the Rashba SOC constant α. When α is small, the band structure is still parabolic. Therefore, researchers usually treat the spin–orbit interaction as a perturbation [4,5,25]. The interaction between conduction electrons and the local magnetic moment can be regarded as a spin–orbit effective field acting on local magnetic moment [17,26]: $\vec{H}_R = \alpha\langle\vec{k}\rangle \times \vec{z}$, where $\langle\vec{k}\rangle$ is the drift of $\vec{k}$ in an external electric field. $\langle\vec{k}\rangle$ is calculated based on the ferromagnetic electron gas model regardless of the band splitting caused by spin–orbit interaction. Consequently, the spin–orbit torque is a direct proportional function of α. When α is large, $\langle\vec{k}\rangle$ is smaller than the result based on the ferromagnetic electron gas model. Our numerical results show the deviation from the proportional relation.

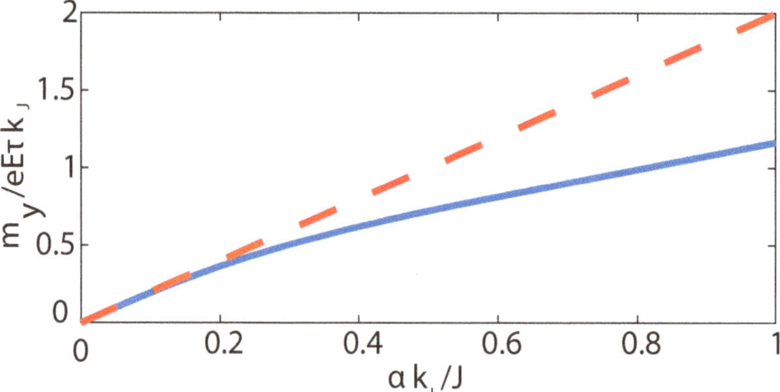

Figure 4. Spin accumulation along the y axis m_y vs. Rashba SOC constant α, where the solid line is the result based on the split bands and the dashed line is based on the band of ferromagnetic electron gas.

Similarly, the spin accumulation along the z axis is

$$m_z = \hbar \int [s_z(\vec{k})(f_\uparrow^0 - f_\downarrow^0)]d\vec{k}.$$

(17)

This spin accumulation depends on the equilibrium distribution function and is parallel to the local magnetic moment; thus, it makes no contribution to the spin–orbit torque.

5. Conclusions

In this paper, we investigated the spin–orbit torque and transport property in a 2D Rashba ferromagnet. The main conclusions are as follows:

The longitudinal conductivity can be divided into two parts: the first term is determined by the charge density and is independent of the spin degrees of freedom. The second term depends on the two bands that spin in opposite directions, and this reflects the spin-polarized transport property of the system.

The spin–orbit torque is directly proportional to the second term of longitudinal conductivity, because spin–orbit torque is just caused by the spin-polarized transport. Although this conclusion is obtained at 0K and using parabolic bands, it is suitable for general linear transport cases. This proportionate relation reveals the underlying connection between spin–orbit torque and conductivity.

Moreover, we demonstrate the impacts of the spin–orbit coupling constant and Fermi energy on spin–orbit torque. When these constants are experimentally changed, the spin–orbit torque can be adjusted accordingly. The results are helpful for relevant experiments.

Author Contributions: Conceptualization, C.Y. and Z.-C.W.; Data curation, C.Y.; Formal analysis, C.Y., D.-K.Z. and Y.-R.W.; Investigation, C.Y., D.-K.Z. and Z.-C.W.; Methodology, D.-K.Z. and Z.-C.W.; Writing—original draft, C.Y.; Writing—review & editing, D.-K.Z., Y.-R.W. and Z.-C.W. All authors have read and agreed to the published version of the manuscript.

Funding: This research was funded by the Natural Science Foundation of Fujian Province, China (Grant No. 2021J05245) and Wuyi University (Grant No. YJ202014). The APC was funded by the Natural Science Foundation of Fujian Province, China (Grant No. 2021J05245).

Data Availability Statement: The data that support the results of this research are available from the corresponding author, [C.Y], upon reasonable request.

Conflicts of Interest: The authors declare no conflict of interest.

References

1. Manchon, A. A new moment for Berry. *Nat. Phys.* **2014**, *10*, 340–341. [CrossRef]
2. Kuschel, T.; Reiss, G. Charges ride the spin wave. *Nat. Nanotechnol.* **2014**, *10*, 22–24. [CrossRef] [PubMed]
3. Molavi, M.; Faizabadi, E. Spin-polarization and spin-flip in a triple-quantum-dot ring by using tunable lateral bias voltage and Rashba spin-orbit interaction. *J. Magn. Magn. Mater.* **2017**, *428*, 488–492. [CrossRef]
4. Manchon, A.; Zhang, S. Theory of nonequilibrium intrinsic spin torque in a single nanomagnet. *Phys. Rev. B* **2008**, *78*, 212405. [CrossRef]
5. Manchon, A.; Zhang, S. Theory of spin torque due to spin-orbit coupling. *Phys. Rev. B* **2009**, *79*, 094422. [CrossRef]
6. Prenat, G.; Jabeur, K.; Pendina, G.D.; Boulle, O.; Gaudin, G. Beyond STT-MRAM, Spin Orbit Torque RAM SOT-MRAM for High Speed and High Reliability Applications. In *Spintronics-Based Computing*; Springer International Publishing: Berlin/Heidelberg, Germany, 2015; pp. 145–157. [CrossRef]
7. Fukami, S.; Zhang, C.; Duttagupta, S.; Kurenkov, A.; Ohno, H. Magnetization switching by spin–orbit torque in an antiferromagnet–ferromagnet bilayer system. *Nat. Mater.* **2016**, *15*, 535–541. [CrossRef]
8. Cubukcu, M.; Boulle, O.; Mikuszeit, N.; Hamelin, C.; Bracher, T.; Lamard, N.; Cyrille, M.C.; Buda-Prejbeanu, L.; Garello, K.; Miron, I.M.; et al. Ultra-Fast Perpendicular Spin–Orbit Torque MRAM. *IEEE Trans. Magn.* **2018**, *54*, 1–4. [CrossRef]
9. Salehi, S.; DeMara, R.F. Adaptive Non-Uniform Compressive Sensing Using SOT-MRAM Multi-Bit Precision Crossbar Arrays. *IEEE Trans. Nanotechnol.* **2021**, *20*, 224–228. [CrossRef]
10. Chung, N.L.; Jalil, M.B.A.; Tan, S.G. Non-equilibrium spatial distribution of Rashba spin torque in ferromagnetic metal layer. *AIP Adv.* **2012**, *2*, 022165. [CrossRef]
11. Wang, X.; Pauyac, C.O.; Manchon, A. Spin-orbit-coupled transport and spin torque in a ferromagnetic heterostructure. *Phys. Rev. B* **2014**, *89*, 054405. [CrossRef]
12. Wang, X.; Manchon, A. Diffusive Spin Dynamics in Ferromagnetic Thin Films with a Rashba Interaction. *Phys. Rev. Lett.* **2012**, *108*, 117201. [CrossRef] [PubMed]
13. Brataas, A.; Kent, A.D.; Ohno, H. Current-induced torques in magnetic materials. *Nat. Mater.* **2012**, *11*, 372–381. [CrossRef] [PubMed]
14. Hals, K.M.D.; Brataas, A. Phenomenology of current-induced spin-orbit torques. *Phys. Rev. B* **2013**, *88*, 085423. [CrossRef]

15. Yang, C.; Wang, Z.C.; Zheng, Q.R.; Su, G. Generalized spin-orbit torques in two-dimensional ferromagnets with spin-orbit coupling. *Eur. Phys. J. B* **2019**, *92*, 1–7. [CrossRef]
16. Jiang, Y.; Jalil, M.B.A. Enhanced spin injection and magnetoconductance by controllable Rashba coupling in a ferromagnet/two-dimensional electron gas structure. *J. Phys. Condens. Matter* **2003**, *15*, L31–L39. [CrossRef]
17. Ado, I.; Qaiumzadeh, A.; Duine, R.; Brataas, A.; Titov, M. Asymmetric and Symmetric Exchange in a Generalized 2D Rashba Ferromagnet. *Phys. Rev. Lett.* **2018**, *121*, 086802. [CrossRef]
18. Haney, P.M.; Lee, H.W.; Lee, K.J.; Manchon, A.; Stiles, M.D. Current-induced torques and interfacial spin-orbit coupling. *Phys. Rev. B* **2013**, *88*, 214417. [CrossRef]
19. Pauyac, C.O.; Chshiev, M.; Manchon, A.; Nikolaev, S.A. Spin Hall and Spin Swapping Torques in Diffusive Ferromagnets. *Phys. Rev. Lett.* **2018**, *120*, 176802. [CrossRef]
20. Zhang, Z.Y. Charge transport through ferromagnet/two-dimensional electron gas/d-wave superconductor junctions with Rashba spin-orbit coupling. *Eur. Phys. J. B* **2008**, *63*, 65–69. [CrossRef]
21. Ado, I.; Dmitriev, I.; Ostrovsky, P.; Titov, M. Anomalous Hall Effect in a 2D Rashba Ferromagnet. *Phys. Rev. Lett.* **2016**, *117*, 046601. [CrossRef]
22. Xiao, D.; Chang, M.C.; Niu, Q. Berry phase effects on electronic properties. *Rev. Mod. Phys.* **2010**, *82*, 1959–2007. [CrossRef]
23. Matos-Abiague, A.; Rodríguez-Suárez, R.L. Spin-orbit coupling mediated spin torque in a single ferromagnetic layer. *Phys. Rev. B* **2009**, *80*, 094424. [CrossRef]
24. Li, H.; Wang, X.; Doğan, F.; Manchon, A. Tailoring spin-orbit torque in diluted magnetic semiconductors. *Appl. Phys. Lett.* **2013**, *102*, 192411. [CrossRef]
25. Pesin, D.A.; MacDonald, A.H. Quantum kinetic theory of current-induced torques in Rashba ferromagnets. *Phys. Rev. B* **2012**, *86*, 014416. [CrossRef]
26. Chernyshov, A.; Overby, M.; Liu, X.; Furdyna, J.K.; Lyanda-Geller, Y.; Rokhinson, L.P. Evidence for reversible control of magnetization in a ferromagnetic material by means of spin–orbit magnetic field. *Nat. Phys.* **2009**, *5*, 656–659. [CrossRef]

materials

Article

The Effect of Grain Size on the Diffusion Efficiency and Microstructure of Sintered Nd-Fe-B Magnets by Tb Grain Boundary Diffusion

Shuai Guo [1,2,3,*], Xiao Yang [2,3], Xiaodong Fan [1,2,3,*], Guangfei Ding [1,2,3], Shuai Cao [1,2], Bo Zheng [1,2], Renjie Chen [1,2,3] and Aru Yan [1,2,3]

[1] CISRI & NIMTE Joint Innovation Center for Rare Earth Permanent Magnets, Ningbo Institute of Material Technology and Engineering, Chinese Academy of Sciences, Ningbo 315201, China; dingguangfei@nimte.ac.cn (G.D.); caoshuai@nimte.ac.cn (S.C.); zhengbo@nimte.ac.cn (B.Z.); chenrj@nimte.ac.cn (R.C.); aruyan@nimte.ac.cn (A.Y.)

[2] Key Laboratory of Magnetic Materials and Devices, Ningbo Institute of Material Technology and Engineering, Chinese Academy of Sciences, Ningbo 315201, China; yangxiaoworkhard@163.com

[3] University of Chinese Academy of Sciences, Beijing 100049, China

* Correspondence: gshuai@nimte.ac.cn (S.G.); fanxiaodong@nimte.ac.cn (X.F.)

Abstract: The grain boundary diffusion process (GBDP) of heavy rare earth Tb is an effective method to improve the coercivity of Nd-Fe-B magnets, and the matrix grain size has a crucial effect on the diffusion efficiency and depth of the Tb element. In this work, magnets with different grain sizes have been fabricated using powder metallurgy to investigate the effect of grain size on Tb diffusion efficiency and the microstructure of Nd-Fe-B-type magnets. After the Tb diffusion process, the coercivity increment of the magnet with 4.9 μm large grain is 8.60 kOe, which is much higher than that of the magnet with 3.0 μm small grain (~5.90 kOe), which clearly demonstrates that the coercivity increment decreases as the grain size decreases. Microstructure analysis suggested that grain refinement significantly increases the total surface area, resulting in narrowing and discontinuity of the grain boundary phase (GBP). Therefore, as the channel for diffusion, the narrowing and discontinuity of the GBP are unfavorable for diffusion, resulting in a decrease in diffusion efficiency.

Keywords: grain size; diffusion efficiency; microstructure; sintered Nd-Fe-B magnets

Citation: Guo, S.; Yang, X.; Fan, X.; Ding, G.; Cao, S.; Zheng, B.; Chen, R.; Yan, A. The Effect of Grain Size on the Diffusion Efficiency and Microstructure of Sintered Nd-Fe-B Magnets by Tb Grain Boundary Diffusion. *Materials* **2022**, *15*, 4987. https://doi.org/10.3390/ma15144987

Academic Editors: Haiou Wang and Dexin Yang

Received: 22 June 2022
Accepted: 13 July 2022
Published: 18 July 2022

Publisher's Note: MDPI stays neutral with regard to jurisdictional claims in published maps and institutional affiliations.

1. Introduction

Nd-Fe-B-based sintered magnets are widely used in many magnetic devices such as traction motors for hybrid or electronic vehicles and wind generators because of their excellent magnetic properties of high remanence and magnetic energy product [1–3]. However, the coercivity of the actual magnet is far below the intrinsic anisotropy field of the $Nd_2Fe_{14}B$ phase (~7.3T), which has become the key factor limiting the application of Nd-Fe-B magnets [2–4]. In order to enhance the coercivity of sintered Nd-Fe-B magnets, the introduction of heavy rare earth (HRE, such as Dy and Tb) elements is an effective method due to the higher anisotropy fields of $HRE_2Fe_{14}B$ phases. However, the resultant reduction in remanence is unavoidable because of the ferrimagnetic coupling between the HRE atoms and Fe atoms. At present, it is widely accepted that the coercivity of sintered Nd-Fe-B magnets is mainly detrimental to the reversal domain wall nucleation mechanism, and the defect in the surface of matrix grains is the main reason that the practical coercivity is lower than the intrinsic anisotropy field of the $Nd_2Fe_{14}B$ phase [5–8]. Accordingly, introducing heavy rare earth (HRE) elements into the surface layer of the matrix phase by the grain boundary diffusion process (GBDP) has been proved as an effective method to enhance the coercivity of sintered Nd-Fe-B magnets with a slight sacrifice of remanence [9–14]. In the GBDP, the typical core–shell structure is formed, and the surface layer of the matrix phase exhibits a higher anisotropy field to suppress the nucleation of the reversal domain, thereby obtaining higher coercivity [15,16].

In addition, it is well known that the coercivity of sintered Nd-Fe-B magnets exhibits a large grain size dependence [17–19]. Grain refinement technologies such as He gas jet milling [20] and the HDDR (hydrogenation disproportionation desorption recombination) process [21] could effectively improve coercivity within a certain range. Furthermore, apart from directly affecting the coercivity of sintered Nd-Fe-B magnets, grain refinement also affects the distribution of the grain boundary phase (GBP). Since the Nd-rich GBP serves as the main channel for HRE elements' diffusion during GBDP, the microstructure of the GBP has a large influence on the diffusion efficiency. Cao et al. [22] reported that higher RE content promoted Tb diffusion and further contributed to the higher coercivity when studying the influence of rare earth content on coercivity. In sintered Nd-Fe-B magnets, the distribution of the GBP is influenced not only by RE content, but also by the grain size. However, there has been scarce research so far on the relationship between grain size and diffusion efficiency. So, it is meaningful work to investigate the joint effect of grain size and the GBDP on coercivity. In this work, original magnets with different grain sizes were designed to systematically investigate the effect on the diffusion efficiency according to microstructure analysis.

2. Materials and Methods

2.1. Experimental Procedure

Commercial strip casting alloys with a nominal composition of $(Pr_{0.2}Nd_{0.8})_{29}Cu_{0.2}Al_{0.05}Co_{0.5}Ga_{0.1}B_{0.98}Fe_{bal}$ were used as the initial alloys. The strip casting alloys were subjected to a subsequent hydrogen decrepitation (HD) process and further jet milling (JM) in a nitrogen atmosphere. The different average particle sizes of 3.0 μm, 2.5 μm, 2.02 μm, and 1.76 μm were controlled by adjusting the parameter of the JM process. $(Nd,Pr)H_x$ powders were prepared by hydrogenating the Nd-Pr alloy consisting of 80 wt.% Nd and 20 wt.% Pr under H_2 pressure of ~200 kPa for 4 hrs at 400 °C. Subsequently, the prepared magnetic powders were mixed with 2 wt.% $(Pr, Nd)H_x$ (Pr:Nd = 2:8, wt.%) powder. Then, the mixed powders were compacted and aligned under a magnetic field of 2.25 T, followed by cold isostatic compacting under a pressure of 150 MPa. The green compacts were each sintered at 1095 °C, 1080 °C, 1065 °C, and 1050 °C for 2 h in a vacuum followed by gas quenching. Then, the as-sintered magnets prepared by different powders were machined into cylinders with a diameter of 10 mm and a height of 4 mm for diffusion. The machined magnets were divided into two groups. The magnets in the test group were immersed in alcohol-based TbH_x suspension for 5 s and then dried in a N_2 atmosphere. Then, the test group and control group were both treated with heat treatment at 900 °C for 2 h and annealed at 500 °C for 2 h.

2.2. Characterization and Analysis Methods

The average particle size of the JM powder was measured by a laser particle size analyzer, HELOS/RODOS-BR. The samples were cut into cylinders at a size of Φ10 mm × 4 mm, and the magnetic properties of the final annealed samples were measured by pulsed field magnetometry (Hirst PFM-14). The microstructures and elemental distribution of the magnets were observed by a scanning electron microscope (SEM) (Quanta FEG 250, FEI Company, Hillsboro, OR, USA) operating at 20 kV. The contents of Pr, Nd, Tb, and Fe in the core and shell of the matrix phase were obtained using energy-dispersive X-ray spectroscopy (EDS). The contents of Tb at different depths in the diffused magnets were detected using a glow discharge atomic emission spectrometer (GD-OES) (Spectruma Analytik GMBH 750HP, Hof, Germany).

3. Results and Discussion

3.1. Grain Size and Magnetic Properties

Figure 1 shows the backscatter electron (BSE) SEM images and corresponding grain size distributions after sintering of magnetic powders with different particle sizes. It is clear that after counting the dimensions of the matrix grains in every figure, the average grain sizes of the sintered magnets, fabricated by JM powders with different average particle sizes, are 4.93 µm, 4.23 µm, 3.18 µm, and 3.00 µm, which are marked as G4.9, G4.2, G3.2, and G3.0 magnets in the discussion below, respectively.

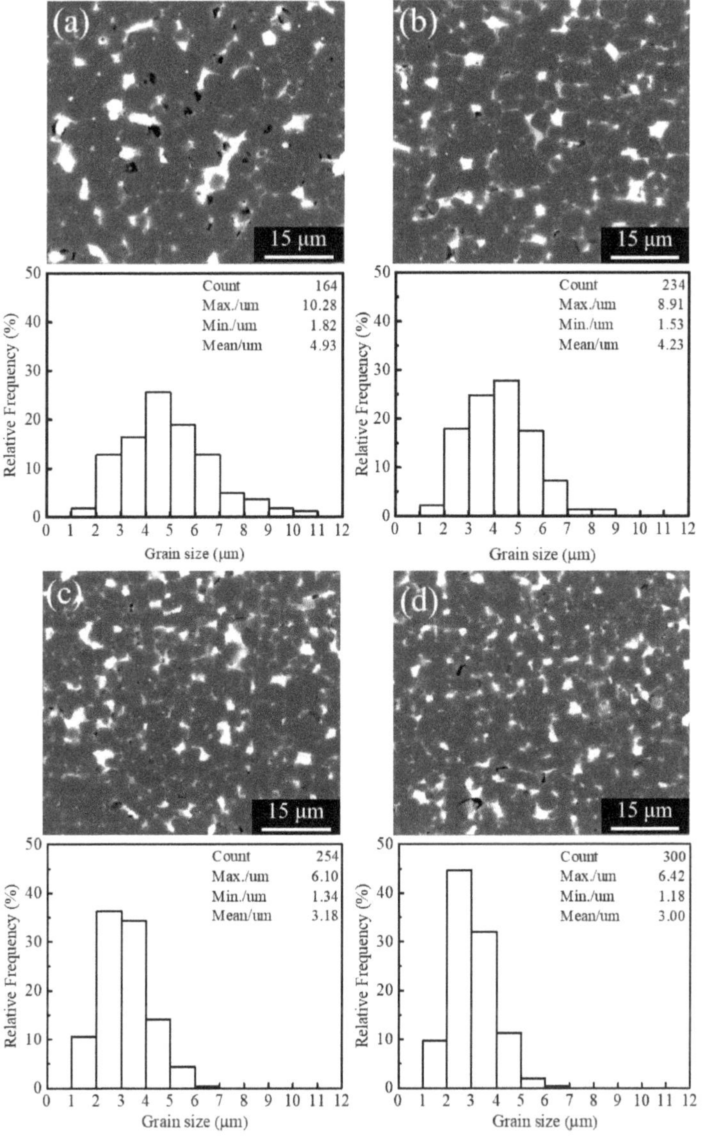

Figure 1. BSE SEM images and corresponding grain size distributions after sintering of magnetic powders with different particle sizes, (**a**) G4.9, (**b**) G4.2, (**c**) G3.2, (**d**) G3.0.

The demagnetization curves of original and diffused magnets with different grain sizes are shown in Figure 2, and the detailed values of coercivity, remanence, and maximum energy product derived from the demagnetization curves are listed in Table 1. The results show that the remanences and energy products of the magnets remain basically the same before and after the diffusion process, but the coercivities of the original magnets increase as the grain size decreases and have been enhanced significantly after diffusion. The more detailed coercivity change after diffusion is shown in Figure 2b, which demonstrates that the coercivity increment (ΔH_{cj} in Figure 2b) after diffusion decreases as the grain size decreases.

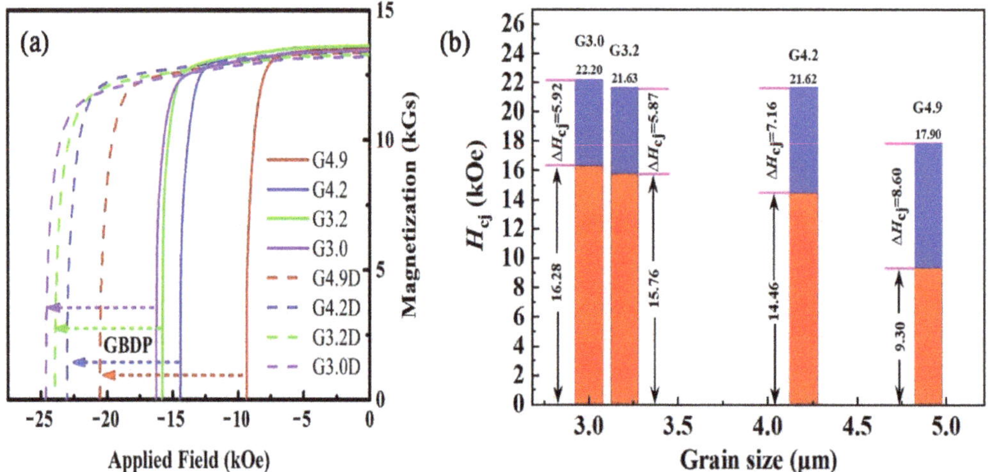

Figure 2. (a) The demagnetization curves of the original magnets (G4.9, G4.2, G3.2, and G3.0) and the diffused magnets (G4.9D, G4.2D, G3.2D, and G3.0D) with different grain sizes; (b) the coercivity of magnets with different grain sizes before and after the diffusion process (the red bars represent H_{cj} for the original magnets, and the blue bars represent the increments of H_{cj} after the diffusion process).

Table 1. Coercivity, remanence, and maximum energy product of the magnets with different grain sizes before and after diffusion.

Sample	H_{cj} (kOe)	B_r (kG)	$(BH)_{max}$ (MGOe)
G4.9	9.30	13.53	44.27
G4.2	14.46	13.64	45.49
G3.2	15.76	13.64	45.47
G3.0	16.28	13.45	44.30
G4.9D	17.90	13.58	45.35
G4.2D	21.62	13.51	44.83
G3.2D	21.63	13.38	44.04
G3.0D	22.20	13.28	43.36

So, it could be concluded that the decrease in grain size is detrimental to diffusion efficiency. In addition, although the original coercivity of the G3.2 magnet is higher than that of the G4.2 magnet, the coercivities of these two magnets after diffusion are almost the same. Therefore, for commercial magnet fabrication, comprehensive consideration of grain refinement technology and the GBDP is necessary. Moreover, since the increase in coercivity of the magnet after the GBDP is derived from the Tb-rich shell structure, the difference in the coercivity increment should be related to the distribution of the shell structure.

3.2. Microstructure and Element Distribution

Figure 3 shows the cross-sectional BSE SEM images of the magnets with different grain sizes after diffusion from the surface to about 150 μm depth. It is clear that a core–shell structure is formed in the matrix grains on the surface layer for the diffused magnets, and there are two phases with different contrast in the matrix grains after diffusing the Tb atoms. The brighter contrast in the shell area of the matrix grain corresponds to the Tb-rich $(Pr,Nd,Tb)_2Fe_{14}B$ phase, while the darker contrast in the core area of the matrix grain corresponds to the Tb-lean $(Pr,Nd)_2Fe_{14}B$ phase. In addition, the thickness of the Tb-rich shell decreases with the increase in diffusion depth for all four different grain size magnets. In the G4.9D magnet (Figure 3a), the Tb-rich shell structure, which is clearly visible in the BSE image surrounding the $(Pr,Nd)_2Fe_{14}B$ core, is well performed. However, when the diffusion depth increases to about 150 μm, the Tb-rich core–shell structure becomes almost invisible. Actually, the core–shell structure still exists when the depth exceeds 150 μm, but the thickness of the Tb-rich shell is too small to see. The most interesting thing is that as the grain size decreases, the depth of the Tb diffusion area with visible core–shell structure decreases gradually. In the G4.2D magnet, the core–shell structure is obvious at a depth of about 100 μm, but in the G3.0D magnet, the apparent core–shell structure can only be found at a depth of less than 50 μm.

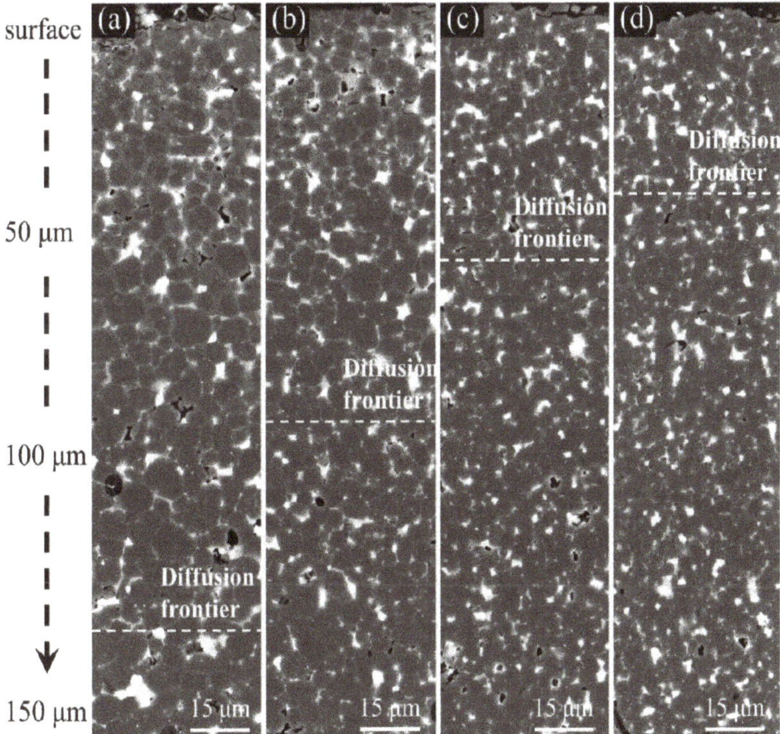

Figure 3. The longitudinal cross-sectional BSE-SEM images of (**a**) G4.9D, (**b**) G4.2D, (**c**) G3.2D, and (**d**) G3.0D diffused magnets.

In order to further investigate accurate Tb content with increasing depth, the concentrations of Tb were obtained for G4.9D, G4.2D, G3.2D, and G3.0D diffused magnets, as shown in Figure 4. Affected by the diffusion dynamics, the distribution of Tb element in all magnets follows the law that the Tb concentration decreases gradually with the increase in diffusion depth. There exist different characteristics of Tb distribution in the magnets due to the different grain sizes. It is obvious that, under the same diffusion depth, Tb concentration in the surface area of the matrix grain for the large grain magnet is higher than that of the small grain magnet. However, as the diffusion depth increases, the concentration of Tb gradually becomes the same in all magnets. Particularly, the Tb concentration in the large grain magnet is still higher than that of the small grain magnet, while the diffusion depth exceeds 150 μm, which means that the large grain size facilitates the infiltration of Tb to the interior of the magnet.

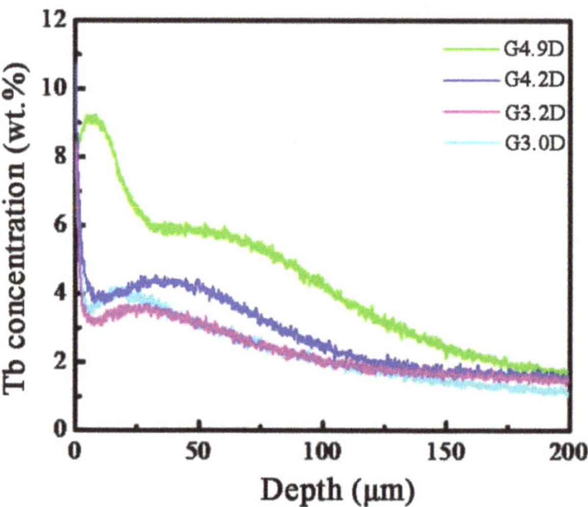

Figure 4. Tb contents as a function of depth from the surface of G4.9D, G4.2D, G3.2D, and G3.0D diffused magnets.

The BSE-SEM images at 15 μm from the surface of magnets are shown in Figure 5. This region is very close to the surface of the magnet, and the concentration of Tb in the grain boundary phase is relatively high. Therefore, the diffusion of Tb to the matrix phase is sufficient and the core–shell structure is obviously formed. However, the thickness of the shell structure in magnets with different grain sizes is not the same. It has been discovered that the thickness of the Tb-rich shell structure decreases with the decrease in grain size. In the G4.9D magnet, Tb has almost penetrated through the entire grain. However, in the G4.2D and G3.2D magnets, there exist distinct shell structures on the matrix grain's surface and the Tb-rich shell in the G4.2D magnet is thicker than that in the G3.2D magnet. However, in the G3.0D magnet, Tb only concentrates in the superficial layer of the matrix phase and the thickness of the Tb-rich shell is much lower than that of the G4.2D and G3.2D magnets.

The contents of Tb in the shell region of the matrix grains in different magnets (points A–D in Figure 5) are shown in Table 2. It is clear that the Tb content in the shell region of the matrix grain gradually decreases with the increasing grain size, which is in accordance with the analysis of the core–shell structure in Figure 5. On the contrary, the contents of Pr and Nd in the shell region increase rapidly as the grain size decreases, which means that a small grain size prevents Tb diffusion to the interior of the magnets. In conclusion, the above results demonstrate that the diffusion depth of Tb atoms in Nd-Fe-B-type magnets has been restricted by the small grain size, and the detailed reason will be discussed below.

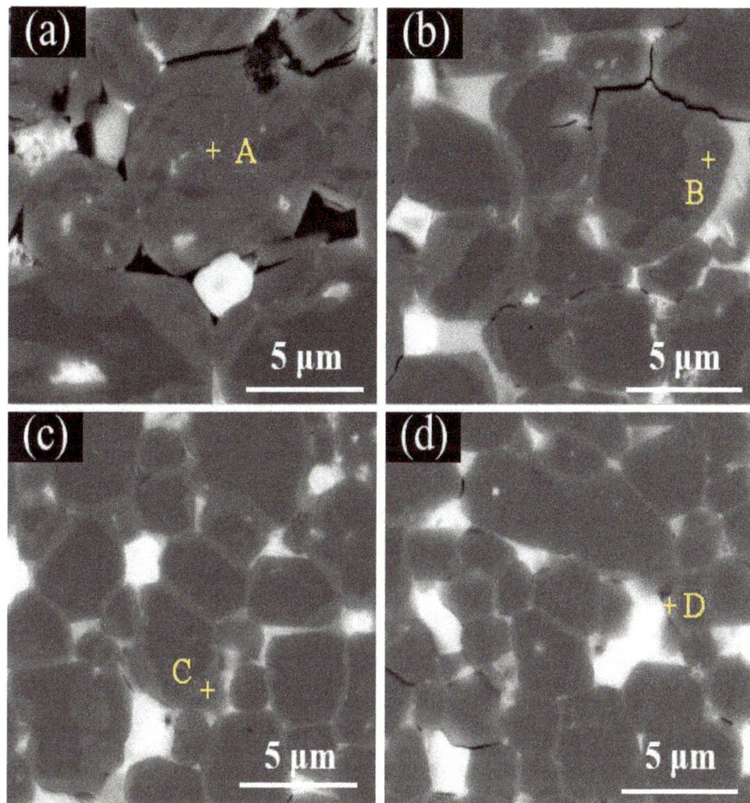

Figure 5. The BSE-SEM images at 15 μm from the surface of (**a**) G4.9D, (**b**) G4.2D, (**c**) G3.2D, and (**d**) G3.0D diffused magnets.

Table 2. The elemental contents (wt.%) of the selected position in Figure 5 using EDS analysis.

Position	Pr	Nd	Tb	Fe
A	2.77	10.90	15.93	70.40
B	4.20	15.75	13.12	66.93
C	4.77	17.50	7.41	70.32
D	6.30	20.07	6.50	67.13

3.3. Diffusion Process Analysis

As is widely known, the GBP is the main channel for Tb diffusion, but the distribution of the GBP is closely related to the grain size. On the one hand, for spherical-like grains, the specific surface area of the single grain increases with the decrease in the grain volume; thus, the total surface area of the matrix grains increases as the grain size decreases. On the other hand, in sintered Nd-Fe-B-type magnets, the volume of GBPs for different magnets is basically the same. Therefore, the volume of GBP covering the unit surface region of the matrix grain surface reduces with the decrease in grain size, which leads to the GBP becoming thinner and more discontinuous. A schematic diagram for the grain boundary diffusion processes of magnets with different grain sizes is shown in Figure 6. The grain size of the G4.9 magnet (R) is larger than the grain size of the G4.2 magnet (r), so the GBP of the G4.9 magnet should be wider and more continuous than the GBP of the G4.2 magnet, that is to say, D > d. Because the GBP is the main channel for the Tb diffusion process, wider and more continuous GBP is beneficial to Tb diffusion [22,23]. In the G4.9 magnet, Tb is easier to diffuse into the interior of magnets along the GBP. With the grain size decreasing, the GBP becomes narrower and more discontinuous, and the grain boundary diffusion of Tb is insufficient; therefore, the diffusion depth is limited. At the same time, due to the Tb concentration in the GBP of the G4.9 magnet being higher than the G4.2 magnet, more Tb atoms diffuse into the main phase of the G4.9 magnet than into the G4.2 magnet. So, the shell thickness of the G4.9D magnet is larger than the G4.2D magnet, which is consistent with the SEM results.

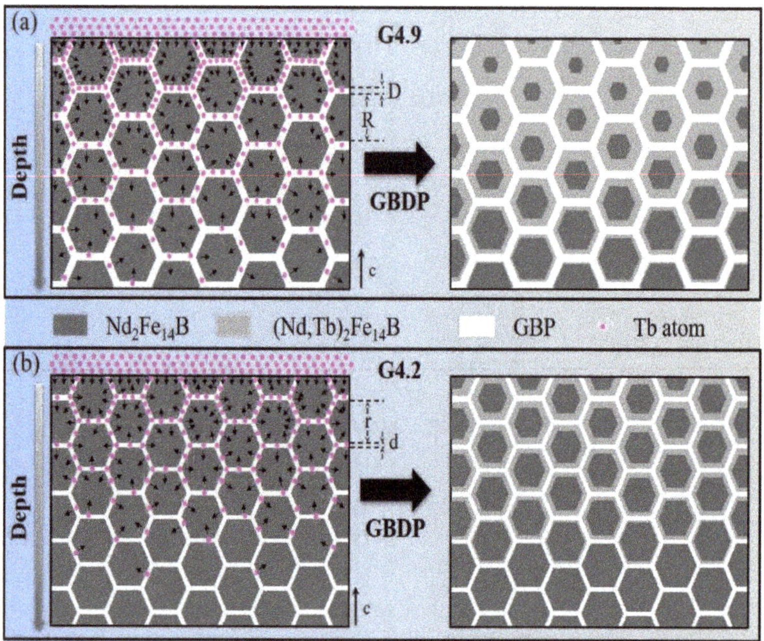

Figure 6. The schematic illustration of the diffusion process of the (**a**) G4.9 and (**b**) G4.2 magnets.

4. Conclusions

In summary, the effect of grain size on Tb diffusion efficiency and microstructure was systematically investigated. After the diffusion process, the coercivity increment decreases as the grain size decreases, and the magnet with an average grain size of 4.9 µm has the largest coercivity increment of 8.60 kOe. Microstructure analysis suggests the variation in the coercivity increment is caused by the difference in grain size. When the grain size decreases, the specific surface area and total surface area of grains gradually increase, which leads to the GBP becoming narrow and discontinuous. Since the GBP is the main channel for Tb diffusion, the narrow and discontinuous GBP inhibits the Tb diffusion. Meanwhile, because of the higher Tb concentration in the GBP of magnets with larger grain sizes, the thickness of the shell structure decreases as the grain size decreases. Finally, the magnets with larger grain sizes have deeper diffusion depth and higher coercivity increments.

Author Contributions: Investigation, formal analysis, writing and editing, S.G. and X.F.; materials preparation and testing, X.Y. and G.D.; testing, writing—review and editing, S.C. and B.Z.; project administration, funding acquisition, R.C. and A.Y. All authors have read and agreed to the published version of the manuscript.

Funding: This research was funded by the Key R&D program of Zhejiang Province (No. 2021C01190), Major Project of "Science and Technology Innovation 2025" in Ningbo City (No. 2020Z046), Inner Mongolia Major Technology Project (2019ZD020), Kunpeng plan of Zhejiang Province and Ningbo top talent program.

Institutional Review Board Statement: Not applicable.

Informed Consent Statement: Not applicable.

Data Availability Statement: The data presented in this study are available on request from the corresponding author.

Conflicts of Interest: The authors declare no conflict of interest.

References

1. Gutfleisch, O.; Willard, M.A.; Brück, E.; Chen, C.H.; Sankar, S.G.; Liu, J.P. Magnetic Materials and Devices for the 21st Century: Stronger, Lighter, and More Energy Efficient. *Adv. Mater.* **2010**, *23*, 821–842. [CrossRef] [PubMed]
2. Wu, Y.; Skokov, K.P.; Schäfer, L.; Maccari, F.; Aubert, A.; Xu, H.; Wu, H.C.; Jiang, C.B.; Gutfleisch, O. Microstructure, coercivity and thermal stability of nanostructured (Nd,Ce)-(Fe,Co)-B hot-compacted permanent magnets. *Acta Mater.* **2022**, *235*, 118062. [CrossRef]
3. Zhao, L.Z.; He, J.Y.; Li, W.; Liu, X.L.; Zhang, J.; Wen, L.; Zhang, Z.H.; Hu, J.W.; Zhang, J.S.; Liao, X.F.; et al. Understanding the role of element grain boundary diffusion mechanism in Nd–Fe–B magnets. *Adv. Funct. Mater.* **2021**, *32*, 2109529. [CrossRef]
4. Hono, K.; Sepehri-Amin, H. Strategy for high-coercivity Nd–Fe–B magnets. *Scr. Mater.* **2012**, *67*, 530–535. [CrossRef]
5. Kronmüller, H. Theory of Nucleation Fields in Inhomogeneous Ferromagnets. *Physica Status Solidi* **1987**, *144*, 385–396. [CrossRef]
6. Sagawa, M.; Fujimura, S.; Togawa, N.; Yamamoto, H.; Matsuura, Y. New material for permanent magnets on a base of Nd and Fe (invited). *J. Appl. Phys.* **1984**, *55*, 2083. [CrossRef]
7. Durst, K.D.; Kronmüller, H. The coercive field of sintered and melt-spun NdFeB magnets. *J. Magn. Magn. Mater.* **1987**, *68*, 63–75. [CrossRef]
8. Lv, M.; Kong, T.; Zhang, W.H.; Zhu, M.Y.; Jin, H.M.; Li, W.X.; Li, Y. Progress on modification of microstructures and magnetic properties of Nd-Fe-B magnets by the grain boundary diffusion engineering. *J. Magn. Magn. Mater.* **2021**, *517*, 167278. [CrossRef]
9. Hirota, K.; Nakamura, H.; Minowa, T.; Honshima, M. Coercivity Enhancement by the Grain Boundary Diffusion Process to Nd–Fe–B Sintered Magnets. *IEEE Trans. Magn.* **2006**, *42*, 2909–2911. [CrossRef]
10. Sepehri-Amin, H.; Ohkubo, T.; Hono, K. Grain boundary structure and chemistry of Dy-diffusion processed Nd–Fe–B sintered magnets. *J. Appl. Phys.* **2010**, *107*, 09A745. [CrossRef]
11. Cao, X.J.; Chen, L.; Guo, S.; Li, X.B.; Yi, P.P.; Yan, A.R.; Yan, G.L. Coercivity enhancement of sintered Nd–Fe–B magnets by efficiently diffusing DyF3 based on electrophoretic deposition. *J. Alloys Compd.* **2015**, *631*, 315–320. [CrossRef]
12. Cao, X.; Chen, L.; Guo, S.; Chen, R.; Yan, G.; Yan, A. Impact of TbF$_3$ diffusion on coercivity and microstructure in sintered Nd–Fe–B magnets by electrophoretic deposition. *Scr. Mater.* **2016**, *116*, 40–43. [CrossRef]
13. Di, J.; Ding, G.; Tang, X.; Yang, X.; Guo, S.; Chen, R.; Yan, A. Highly efficient Tb-utilization in sintered Nd-Fe-B magnets by Al aided TbH$_2$ grain boundary diffusion. *Scr. Mater.* **2018**, *155*, 50–53. [CrossRef]
14. Liu, Z.W.; He, J.Y.; Ramanujan, R.V. Significant progress of grain boundary diffusion process for cost effective rare earth permanent magnets: A review. *Mater. Des.* **2021**, *209*, 110004. [CrossRef]
15. Kim, T.H.; Sasaki, T.T.; Koyama, T.; Fujikawa, Y.; Miwa, M.; Enokido, Y.; Ohkubo, T.; Hono, K. Formation mechanism of Tb-rich shell in grain boundary diffusion processed Nd–Fe–B sintered magnets. *Scripta Mater.* **2020**, *178*, 433–437. [CrossRef]
16. Lu, K.C.; Bao, X.Q.; Song, X.F.; Wang, Y.X.; Tang, M.H.; Li, J.H.; Gao, X.X. Temperature dependences of interface reactions and Tb diffusion behavior of Pr-Tb-Cu-Al alloys/Nd-Fe-B magnet. *Scripta Mater.* **2021**, *191*, 90–95. [CrossRef]
17. Ramesh, G.R.; Thomas, B.M. Magnetization reversal in nucleation controlled magnets. II. Effect of grain size and size distribution on intrinsic coercivity of Fe-Nd-B magnets. *J. Appl. Phys.* **1998**, *64*, 6416–6423. [CrossRef]
18. Uestuener, K.; Katter, M.; Rodewald, W. Dependence of the Mean Grain Size and Coercivity of Sintered Nd-Fe-B Magnets on the Initial Powder Particle Size. *IEEE Trans. Magn.* **2006**, *42*, 2897–2899. [CrossRef]
19. Sepehri-Amin, H.; Ohkubo, T.; Gruber, M.; Schreflb, T.; Hono, K. Micromagnetic simulations on the grain size dependence of coercivity in anisotropic Nd–Fe–B sintered magnets. *Scr. Mater.* **2014**, *89*, 29–32. [CrossRef]
20. Sepehri-Amin, H.; Une, Y.; Ohkubo, T.; Hono, K.; Sagawa, M. Microstructure of fine-grained Nd–Fe–B sintered magnets with high coercivity. *Scr. Mater.* **2011**, *65*, 396–399. [CrossRef]
21. Ding, G.; Guo, S.; Chen, L.; Di, J.; Chen, K.; Chen, R.; Lee, D.; Yan, A. Effects of the grain size on domain structure and thermal stability of sintered Nd-Fe-B magnets. *J. Alloys Compd.* **2018**, *735*, 1176–1180. [CrossRef]
22. Cao, X.; Chen, L.; Guo, S.; Fan, F.; Chen, R.; Yan, A. Effect of rare earth content on TbF$_3$ diffusion in sintered Nd–Fe–B magnets by electrophoretic deposition. *Scr. Mater.* **2017**, *131*, 24–28. [CrossRef]
23. Xu, F.; Wang, J.; Dong, X.; Zhang, L.; Wu, J. Grain boundary microstructure in DyF$_3$-diffusion processed Nd–Fe–B sintered magnets. *J. Alloys Compd.* **2011**, *509*, 7909–7914. [CrossRef]

reference [45], the ground-state phase transition and magnetic properties of a frustrated spin chain with side chains have been investigated in detail. Motivated by novel, highly correlated low-dimensional systems, V. O. Cheranovskii et al. studied the magnetization process of a series of spin-1/2 chains with different types of intra-chain interactions at low temperatures [46]. The Kondo-necklace models with similar spin structure to the branched chains were considered, and some important results for their ground-state phase transition have been reported [47,48].

Despite the fact that linear spin chains with side spins comprise a rich physics in materials science and technology, less attention has been paid to the effect of the number of branches, to the nature of interactions between chain and side spins, and to the effect of intra-chain interactions on the ground-state phase spectra of these models. In the present paper, we will numerically discuss the low-temperature magnetic properties of the spin-1/2 XXX Heisenberg model on the three different branched chains with one, two and three intra-chain interactions in their unit blocks. Hereafter, we label these chains as chain Y_1, chain Y_2 and chain Y_3, respectively. In fact, we will widely discuss the problem concerning the effects of side spins and their exchange interaction on the ground-state phase spectra and the magnetization process of the above-described spin-1/2 Heisenberg branched chains. To gain a reasonable insight into the low-temperature magnetization process of the chain models, we use Quantum Monte Carlo (QMC) simulations under the subroutine dirloop–sse–a package from the Algorithms and Libraries for Physics Simulations (ALPS) project, which prepares a full generic implementation of the QMC simulations for spin lattices [49,50].

The paper is organized as follows. In Section 2, the spin-1/2 XXX Heisenberg model on the three different types of 1-D branched chains with one, two and three intra-chain interactions per unit cell is introduced. In Section 3, we numerically investigate the magnetic properties of the introduced spin chains by implementing QMC simulations. The following input parameters are considered for the applied sse loop: 1×10^5 thermalization, 1×10^4 MC sweeps, assuming periodic boundary conditions. Next, we report the most interesting results obtained for the magnetic behavior of the spin chains Y_1, Y_2 and Y_3 under two different conditions considered for the nature of inter- and intra-chain interactions J_1 and J_2. In particular, the effect of the intra-chain interaction J_2 between side spins and chains on the low-temperature magnetization of the introduced models will be discussed in detail. The conclusions are drawn in Section 4.

2. The Model

Let us consider the spin-1/2 XXX Heisenberg model on the branched chains illustrated in Figure 1 described by an effective Hamiltonian

$$\mathcal{H} = \sum_{i=1}^{N} \left\{ J_1 \left(S_{A_i} \cdot S_{B_i} + S_{B_i} \cdot S_{A_{i+1}} \right) + \sum_{j=1}^{n} \left[J_2 \left(S_{A_i} \cdot S_{C_{i,j}} \right) - g\mu_B H \left(S_{A_i}^z + S_{B_i}^z + S_{C_{i,j}}^z \right) \right] \right\} \tag{1}$$

where N is the number of unit cells, and n accounts for the number of side spins. The total number of spins $L = (n+2)N$ is considered for the systems under periodic boundary conditions, where $S_{A_{N+1}} \equiv S_{A_1}$. J_1 and J_2 are, respectively, inter- and intra-chain exchange interactions between each of the two nearest neighbor spins, H is the applied magnetic field in the z–direction. g and μ_B are g-factor and Bohr magneton, respectively. The XXX interaction between each pair of spins can be formulated as

$$S_\alpha \cdot S_{\alpha'} = S_\alpha^x S_{\alpha'}^x + S_\alpha^y S_{\alpha'}^y + S_\alpha^z S_{\alpha'}^z \tag{2}$$

in which spin-1/2 operators are given by ($\hbar = 1$)

$$s^x = \frac{1}{2} \begin{pmatrix} 0 & 1 \\ 1 & 0 \end{pmatrix}, \quad s^y = \frac{1}{2} \begin{pmatrix} 0 & -i \\ i & 0 \end{pmatrix}, \quad s^z = \frac{1}{2} \begin{pmatrix} 1 & 0 \\ 0 & -1 \end{pmatrix} \tag{3}$$

Henceforward, to have a dimensionless parameter space, in our numerical tasks, the intra-chain exchange interaction J_2 will be considered as the energy unit.

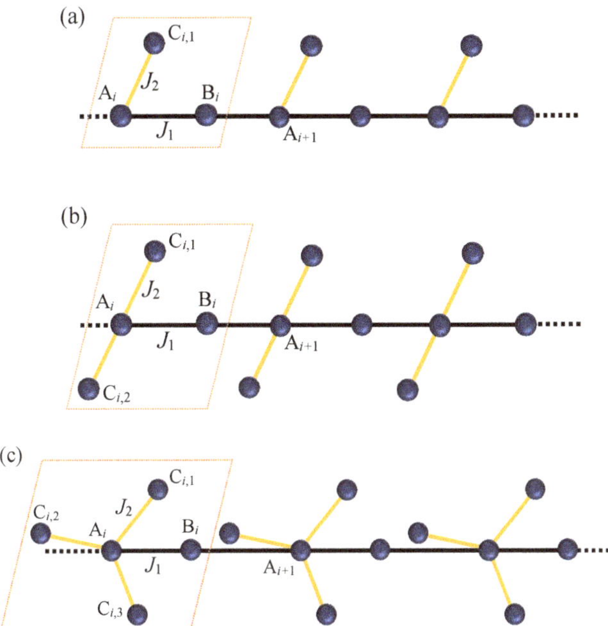

Figure 1. Schematic structure of the spin-1/2 XXX Heisenberg model on the branched chains (**a**) Y_1 with one, (**b**) Y_2 with two, and (**c**) Y_3 with three intra-chain interactions J_2. J_1 denotes inter-chain interaction. In each case, the dotted rectangle indicates a unit cell that uniformly repeats throughout the chains. Solid balls labeled as A, B, and C represent spin-1/2 particles in a unit block, and the balls marked with C_i indicate the side spins.

3. Magnetic Properties

To demonstrate the magnetic properties of the three types of 1-D spin-1/2 Heisenberg branched chains Y_{1-3}, we have calculated the low-temperature magnetization process using the QMC method with the stochastic series expansion implementation from the ALPS package with $L = 120$ number of spins. Two following conditions are supposed for the exchange couplings J_1 and J_2:

❖ Condition (I): We assume ferromagnetic coupling $|J_2|$ is the energy unit ($J_2 < 0$) where the magnetization of the models is examined for different antiferromagnetic interaction ratios $J_1/|J_2| > 0$.

❖ Condition (II): We assume antiferromagnetic coupling $J_2 > 0$ is the energy unit, and the magnetization is investigated for various fixed values of the ferromagnetic interaction ratio $J_1/J_2 < 0$.

The main goal of the above selection of antiferromagnetic–ferromagnetic interactions is to study the competition between antiferromagnetism and ferromagnetism to determine the predomination of the gapless spin liquid phases and to detect the Kosterlitz–Thouless (KT) critical point in the ground-state phase spectra of the models.

3.1. The Heisenberg Spin-1/2 Branched Chains under Condition (I)

Figure 2 displays QMC simulations of the magnetization per saturation value M/M_s for the chains Y_{1-3} versus the magnetic field at low enough temperature $k_B T/|J_2| = 0.02$ (T is the temperature and k_B is the Boltzmann constant).

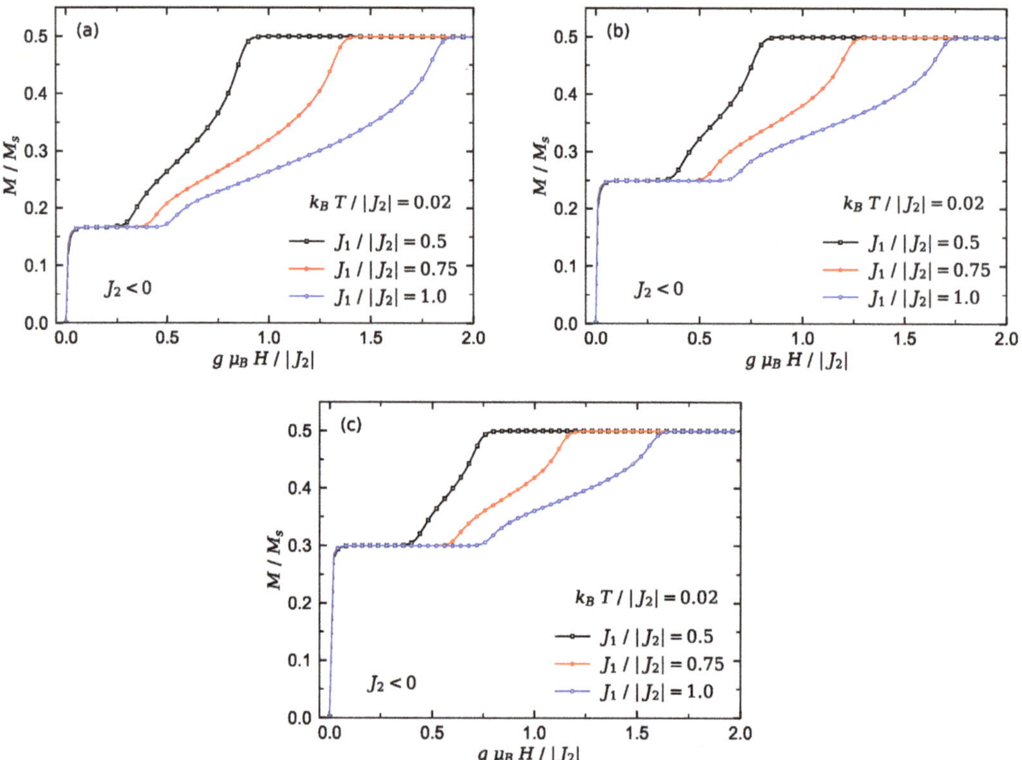

Figure 2. QMC results obtained for the magnetization curve of the three Heisenberg branched chains at low temperatures $k_B T/|J_2| = 0.02$ under the condition (I) where three different fixed values of the inter-chain interaction $J_1/|J_2| = \{0.5, 0.75, 1.0\}$ are considered. (**a**) The magnetization of chain Y_1. (**b**) The magnetization of chain Y_2. (**c**) The magnetization of chain Y_3.

As can be seen in Figure 2, the magnetization of each spin chain shows an abrupt jump when the magnetic field increases from zero. This quantity reaches an intermediate magnetization plateau at the fractional value $M/M_s = n/(n+2)$. This single plateau coincides with the gapful Lieb–Mattis (LM) ferrimagnetic ground state [25,51]. Obviously, increasing the number of side spins n results in enhancing the height and width of LM plateau. The intermediate plateau terminates at a critical magnetic field due to a quantum phase transition to a Luttinger spin liquid phase. Evidently, increasing the interaction ratio $J_1/|J_2|$ enhances the region of the spin liquid phase, as well as the saturation field, while increasing the number of side spins restricts the region of the spin liquid phase and decreases the saturation field (compare Figure 2a with Figure 2b,c).

3.2. The Heisenberg Spin-1/2 Branched Chains under Condition (II)

The spin chains Y_2 and Y_3 show more interesting magnetic properties under the condition (II). To shed light on this issue, we plot the magnetization curve of models Y_{1-3} versus the magnetic field at $k_B T / J_2 = 0.02$, where a few selected values of the interaction ratio J_1 / J_2 are supposed. The model Y_1 displays in its magnetization process a single intermediate plateau at one-third of the saturation magnetization (see Figure 3a). When the interaction ratio J_1 / J_2 becomes ferromagnetically stronger, a spin liquid phase emerges into the ground-state phase spectra of this model.

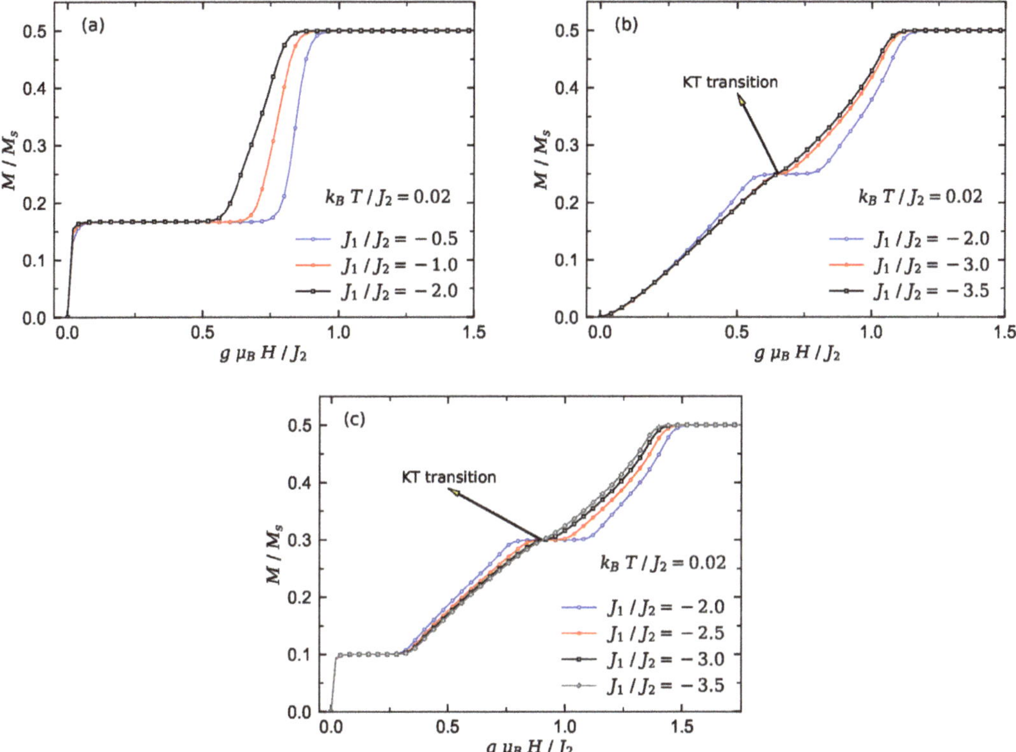

Figure 3. QMC results obtained for the magnetization curve of the three Heisenberg branched chains at low temperatures $k_B T / J_2 = 0.02$ under the condition (II), assuming a few fixed values of the inter-chain interaction J_1 / J_2. (**a**) Chain Y_1. (**b**) Chain Y_2. (**c**) Chain Y_3.

By inspecting Figure 3b, we realize that except for the case shown in Figure 2b, the magnetization of the model Y_2 under condition (II) shows a wide spin liquid phase at low magnetic fields instead of the LM ferrimagnetic phase. An intermediate magnetization plateau at one-half of the saturation value appears at moderate magnetic fields. We find that this intermediate magnetization plateau terminates at a quantum KT critical point when the absolute value of interaction ratio J_1 / J_2 increases.

The sharp decrease in the susceptibility χ of chain Y_2 (see Figure 4a) nearby the relevant quantum critical point $g\mu_B H / J_2 \approx 0.65$ is solid evidence of the KT transition.

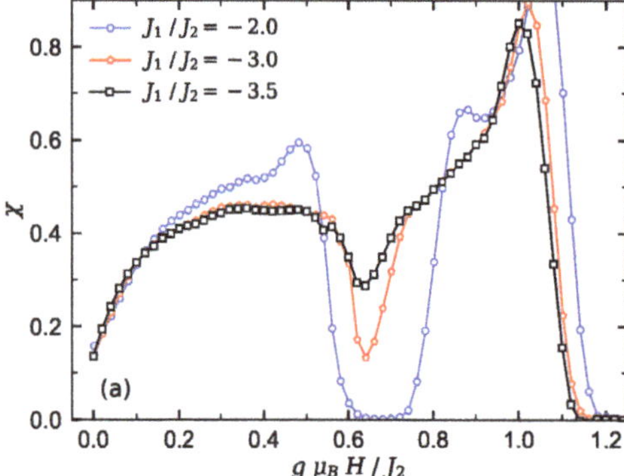
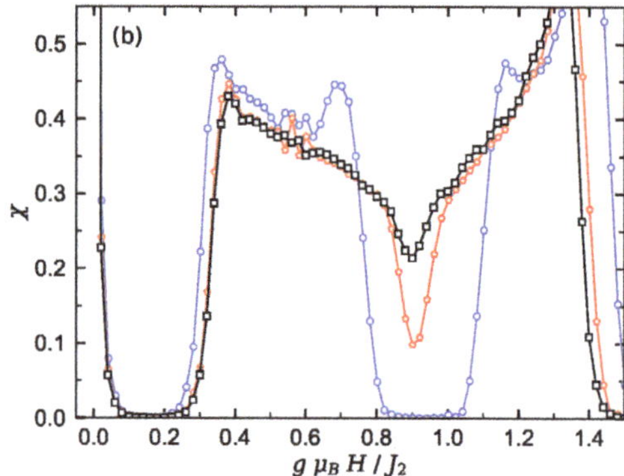

Figure 4. The low-temperature susceptibility χ versus magnetic field for (**a**) the Heisenberg branched chain Y_2 and (**b**) chain Y_3, at $k_B T / J_2 = 0.02$ under the condition (II). Three different values of the inter-chain interaction $J_1 / J_2 = \{-2.0, -3.0, -3.5\}$ are considered for the both panels.

When the number of side spins increases, the branched chain illustrates different magnetization behavior at low temperature. For instance, in Figure 3c, the magnetization of chain Y_3 is presented for different fixed values of the interaction ratio J_1 / J_2. It is quite obvious that the magnetization of model Y_3 manifests two intermediate plateaus at one-fifth and three-fifth of the saturation magnetization. In fact, by increasing the magnetic field from zero, the magnetization abruptly jumps to the first intermediate plateau. This quantity starts to increase smoothly from the first plateau at a critical magnetic field $(g\mu_B H / J_2 \approx 0.3)$ and reaches the second intermediate plateau, and successively a second smooth increase of the magnetization starts from another critical magnetic field nearby $g\mu_B H / J_2 \approx 1.1$. These smooth increases of the magnetization are reminiscent of the existence of the Tomonaga–Luttinger spin liquid in the ground-state phase diagram of chain Y_3. It can be seen from Figure 3c that the second intermediate plateau $(M / M_s = 3/5)$ appeared in the magnetization curve of the branched chain Y_3 monotonically shrinks upon increasing the interaction ratio J_1 / J_2. This plateau eventually disappears at a quantum critical KT point $g\mu_B H / J_2 \approx 0.9$. The steep decrease in the susceptibility χ shown in Figure 4b close to this point denotes a quantum KT transition. Clearly, magnetic susceptibility χ vanishes for the gapful phases, while it has unconventional behavior with nonzero value within the gapless spin liquid phase regions.

4. Conclusions

To summarize, we have numerically investigated the magnetic properties of the spin-1/2 XXX Heisenberg model on the three different 1-D branched chains consisting of one, two and three intra-chain interactions per unit cell. Indeed, we have examined the magnetization and magnetic susceptibility of these spin systems by employing the QMC method.

When the intra-chain interaction is ferromagnetic and inter-chain interaction is antiferromagnetic, the magnetization of all chains shows a single intermediate plateau whose height is related to the number of side spins. This plateau terminates at a critical field, and successively, a gapless Tomonaga–Luttinger liquid phase arises, where the magnetization starts to increase smoothly until it reaches its saturation value. We understood that by increasing the number of side spins, the phase region corresponds to the Luttinger liquid phase diminishes.

On the other hand, when the intra-chain interaction is antiferromagnetic and inter-chain interaction is ferromagnetic, the low-temperature magnetization of the branched chains with more than one side spin per unit cell exhibit more rich magnetic properties. Under this situation, the magnetization of the branched chain with two side spins shows an intermediate plateau at one-half of saturation value that breaks the quantum spin liquid into two areas. We have concluded that for a relatively strong value of the ferromagnetic inter-chain interaction this plateau terminates at a quantum KT critical point.

Moreover, we have observed that the magnetization of chain with three side spins shows two intermediate plateaus at one-fifth and three-fifth of the saturation magnetization. Another interesting finding from our investigations is that with an increase in the inter-chain interaction, the intermediate plateau at three-fifth of saturation magnetization gradually shrinks and eventually ends up at a KT point, while the intermediate plateau at one-fifth of saturation magnetization becomes more robust.

We theoretically concluded that the considered Heisenberg branched chains reveal an interesting magnetic response to the quantity and quality of the intra-chain interaction. It is expected that our results could be useful for explaining the magnetic behavior of various metal-containing polymers with different side spins.

Author Contributions: Data curation, H.A.Z.; Formal analysis, A.Z.; Investigation, H.A.Z. and N.A.; Methodology, A.Z.; Project administration, H.A.Z.; Resources, A.Z.; Software, H.A.Z. and A.Z.; Supervision, N.A. and M.J.; Validation, M.J.; Visualization, A.Z.; Writing—original draft, H.A.Z.; Writing—review and editing, H.A.Z., A.Z. and M.J. All authors have read and agreed to the published version of the manuscript.

Funding: This work was supported by the SAIA (Slovak Academic Information Agency), and Ministry of Education, Science,Research and Sport of the Slovak Republic under contract No. VEGA 1/0531/19 and the Slovak Research and Development Agency provided under Contract No. APVV-20-0150.

Acknowledgments: H. Arian Zad and A. Zoshki acknowledge the financial support of the National Scholarship Programme of the Slovak Republic (NŠP). H. Arian Zad also acknowledges the receipt of the grant from the Abdus Salam International Centre for Theoretical Physics (ICTP), Trieste, Italy. N. Ananikian acknowledges the support of CS MES RA in the frame of the research project No. SCS 19IT-008 and research project No. SCS 21AG-1C006. M. Jaščur has been supported by the Grant of Ministry of Education, Science, Research and Sport of the Slovak Republic under contract No. VEGA 1/0531/19, and Grant of the Slovak Research and Development Agency provided under Contract No. APVV-20-0150.

Conflicts of Interest: The authors declare no conflict of interest.

References

1. Kuramoto, T. Magnetic and Critical Properties of Alternating Spin Heisenberg Chain in a Magnetic Field. *J. Phys. Soc. Jpn.* **1998**, *67*, 1762. [CrossRef]
2. Yamanoto, S.; Sakai, T. Breakdown of a magnetization plateau due to anisotropy in Heisenberg mixed-spin chains. *J. Phys. Condens. Matter* **1999**, *11*, 5175. [CrossRef]
3. Sakai, T.; Yamamoto, S. Critical behavior of anisotropic Heisenberg mixed-spin chains in a field. *Phys. Rev. B* **1999**, *60*, 4053. [CrossRef]
4. Ivanov, N.B. Magnon dispersions in quantum Heisenberg ferrimagnetic chains at zero temperature. *Phys. Rev. B* **2000**, *62*, 3271. [CrossRef]
5. Rojas, M.; Rojas, O.; de Souza, S.M. Frustrated Ising model on the Cairo pentagonal lattice. *Phys. Rev. E* **2012**, *86*, 051116. [CrossRef] [PubMed]
6. Rousochatzakis, I.; Läuchli, A.M.; Moessner, R. Quantum magnetism on the Cairo pentagonal lattice. *Phys. Rev. B* **2012**, *85*, 104415. [CrossRef]
7. Rodrigues, F.C.; de Souza, S.M.; Rojas, O. Geometrically frustrated Cairo pentagonal lattice stripe with Ising and Heisenberg exchange interactions. *Ann. Phys.* **2017**, *379*, 1. [CrossRef]
8. Mermin, N.D.; Wagner, H. Absence of Ferromagnetism or Antiferromagnetism in One- or Two-Dimensional Isotropic Heisenberg Models. *Phys. Rev. Lett.* **1966**, *17*, 1133. [CrossRef]
9. Halperin, B.E. On the Hohenberg–Mermin–Wagner Theorem and Its Limitations. *J. Stat. Phys.* **2018**, *175*, 521. [CrossRef]
10. Hohenberg, P.C. Existence of Long-Range Order in One and Two Dimensions. *Phys. Rev.* **1967**, *158*, 383. [CrossRef]
11. Zivieri, R. Absence of Spontaneous Spin Symmetry Breaking in 1D and 2D Quantum Ferromagnetic Systems with Bilinear and Biquadratic Exchange Interactions. *Symmetry* **2020**, *12*, 2061. [CrossRef]
12. Bonner, J.C.; Fisher, M.E. Linear Magnetic Chains with Anisotropic Coupling. *Phys. Rev.* **1964**, *135*, A640. [CrossRef]
13. Carlin, R.L. *Magnetochemistry*; Springer: Dordrecht, The Netherlands, 1986.

14. Kahn, O. *Molecular Magnetism*, 1st ed.; Wiley-VCH: New York, NY, USA, 1993.

15. Torrico, J.; Ohanyan, V.; Rojas, O. Non-conserved magnetization operator and 'fire-and-ice' ground states in the Ising-Heisenberg diamond chain. *J. Magn. Magn. Mater.* **2018**, *454*, 85–96. [CrossRef]

16. Chen, W.P.; Singleton, J.; Qin, L.; Camón, A.; Engelhardt, L.; Luis, F.; Winpenny, R.E.; Zheng, Y.Z. Quantum Monte Carlo simulations of a giant $\{Ni_{21}Gd_{20}\}$ cage with a S=91 spin ground state. *Nat. Comm.* **2018**, *9*, 2107. [CrossRef] [PubMed]

17. Eggert, S.; Affleck, I.; Takahashi, M. Susceptibility of the spin 1/2 Heisenberg antiferromagnetic chain. *Phys. Rev. Lett.* **1994**, *73*, 332. [CrossRef] [PubMed]

18. Eggert, S. Accurate determination of the exchange constant in Sr_2CuO_3 from recent theoretical results. *Phys. Rev. B* **1996**, *53*, 5116. [CrossRef] [PubMed]

19. Pardo, E.; Ruiz-Garciá, R.; Lloret, F.; Faus, J.; Julve, M.; Journaux, Y.; Delgado, F.; Ruiz-Peréz, C. Cobalt(II)–Copper(II) Bimetallic Chains as a New Class of Single-Chain Magnets. *Adv. Mater.* **2004**, *16*, 1597. [CrossRef]

20. Choi, S.W.; Kwak, H.Y.; Yoon, J.H.; Kim, H.C.; Koh, E.K.; Hong, C.S. Intermolecular Contact-Tuned Magnetic Nature in One-Dimensional 3d–5d Bimetallic Systems: From a Metamagnet to a Single-Chain Magnet. *Inorg. Chem.* **2008**, *47*, 10214. [CrossRef]

21. Yamamoto, S. Magnetic Properties of Quantum Ferrimagnetic Spin Chains. *Phys. Rev. B* **1999**, *59*, 1024. [CrossRef]

22. Sakai, T.; Yamamoto, S. Quantum magnetization plateaux in ferrimagnetic spin chains. *J. Magn. Magn. Mater.* **2001**, *226*, 645. [CrossRef]

23. Strečka, J.; Verkholyak, T. Magnetic Signatures of Quantum Critical Points of the Ferrimagnetic Mixed Spin-(1/2, S) Heisenberg Chains at Finite Temperatures. *J. Low Temp. Phys.* **2017**, *187*, 712. [CrossRef]

24. Veríssimo, L.M.; Pereira, M.S.S.; Strečka, J.; Lyra, M.L. Kosterlitz-Thouless and Gaussian criticalities in a mixed spin-($\frac{1}{2}, \frac{5}{2}, \frac{1}{2}$) Heisenberg branched chain with exchange anisotropy. *Phys. Rev. B* **2019**, *99*, 134408. [CrossRef]

25. Strečka, J. Breakdown of a Magnetization Plateau in Ferrimagnetic Mixed Spin-(1/2,S) Heisenberg Chains due to a Quantum Phase Transition towards the Luttinger Spin Liquid. *Act. Phys. Pol. A* **2017**, *131*, 624. [CrossRef]

26. Motoyama, N.; Eisaki, H.; Uchida, S. Magnetic Susceptibility of Ideal Spin 1/2 Heisenberg Antiferromagnetic Chain Systems, Sr_2CuO_3 and $SrCuO_2$. *Phys. Rev. Lett.* **1996**, *76*, 3212. [CrossRef] [PubMed]

27. Oshikawa, M.; Ueda, K.; Aoki, H.; Ochiai, A.; Kohgi, M. Field-Induced Gap Formation in Yb_4As_3. *J. Phys. Soc. Jpn.* **1999**, *68*, 3181. [CrossRef]

28. Oshikawa, M.; Affleck, I. Field-Induced Gap in S = 1/2 Antiferromagnetic Chains. *Phys. Rev. Lett.* **1997**, *79*, 2883. [CrossRef]

29. Kikuchi, H.; Fujii, Y.; Chiba, M.; Mitsudo, S.; Idehara, T.; Tonegawa, T.; Okamoto, K.; Sakai, T.; Kuwai, T.; Ohta, H. Experimental observation of the 1/3 magnetization plateau in the diamond-chain compound $Cu_3(CO_3)_2(OH)_2$. *Phys. Rev. Lett.* **2005**, *94*, 227201. [CrossRef]

30. Hida, K. Magnetic Properties of the Spin-1/2 Ferromagnetic-Ferromagnetic-Antiferromagnetic Trimerized Heisenberg Chain. *J. Phys. Soc. Jpn.* **1994**, *63*, 2359–2364. [CrossRef]

31. Oshikawa, M.; Yamanaka, M.; Affleck, I. Magnetization Plateaus in Spin Chains: "Haldane Gap" for Half-Integer Spins. *Phys. Rev. Lett.* **1997**, *78*, 1984. [CrossRef]

32. Leiner, J.C.; Oh, J.; Kolesnikov, A.I.; Stone, M.B.; Le, M.D.; Kenny, E.P.; Powell, B.J.; Mourigal, M.; Gordon, E.E.; Whangbo, M.-H.; et al. Magnetic excitations of the Cu^{2+} quantum spin chain in Sr_3CuPtO_6. *Phys. Rev. B* **2018**, *97*, 104426. [CrossRef]

33. Arian Zad, H.; Ananikian, N. Phase transitions and magnetization of the mixed-spin Ising-Heisenberg double sawtooth frustrated ladder *J. Phys. Condens. Matter.* **2017**, *29*, 455402. [CrossRef]

34. Arian Zad, H.; Ananikian, N.; Jaščur, M. Single-ion anisotropy effects on the demagnetization process of the alternating weak-rung interacting mixed spin-(1/2, 1) Ising-Heisenberg double saw-tooth ladders. *Phys. Scr.* **2020**, *95*, 095702. [CrossRef]

35. Arian Zad, H.; Trombettoni, A.; Ananikian, N. Spin-1/2 Ising–Heisenberg Cairo pentagonal model in the presence of an external magnetic field: Effect of Landé g-factors. *Eur. Phys. J. B* **2020**, *93*, 200. [CrossRef]

36. Arian Zad, H.; Kenna, R.; Ananikian, N. Magnetic and thermodynamic properties of the octanuclear nickel phosphonate-based cage. *Phys. A* **2020**, *538*, 122841. [CrossRef]

37. Strečka, J.; Karľová, K.; Madaras, T. Giant magnetocaloric effect, magnetization plateaus and jumps of the regular Ising polyhedra. *Phys. B* **2015**, *466–467*, 76–85. [CrossRef]

38. Gálisová, L.; Strečka, J. Magnetic and magnetocaloric properties of the exactly solvable mixed-spin Ising model on a decorated triangular lattice in a magnetic field. *Phys. E* **2018**, *99*, 244. [CrossRef]

39. Strečka, J.; Richter, J.; Derzhko, O.; Verkholyak, T.; Karľová, K. Magnetization process and low-temperature thermodynamics of a spin-1/2 Heisenberg octahedral chain. *Phys. B* **2018**, *536*, 364. [CrossRef]

40. Strečka, J.; Karľová, K; Baliha, V.; Derzhko, O. Ising versus Potts criticality in low-temperature magnetothermodynamics of a frustrated spin-$\frac{1}{2}$ Heisenberg triangular bilayer. *Phys. Rev. B* **2018**, *98*, 174426. [CrossRef]

41. Balents, L. Spin liquids in frustrated magnets. *Nature* **2010**, *464*, 199. [CrossRef] [PubMed]

42. Savary, L.; Balents, L. Quantum spin liquids: A review. *Rep. Prog. Phys.* **2017**, *80*, 016502. [CrossRef] [PubMed]

43. Karľová, K.; Strečka, J.; Lyra, M.L. Breakdown of intermediate one-half magnetization plateau of spin-1/2 Ising-Heisenberg and Heisenberg branched chains at triple and Kosterlitz-Thouless critical points. *Phys. Rev. E* **2019**, *100*, 042127. [CrossRef] [PubMed]

44. Jiang, J.J. Ground-state properties of a quantum frustrated spin chain with side spins and arbitrary spin s. *J. Mag. Magn. Mat.* **2021**, *539*, 168392. [CrossRef]

45. Takano, K.; Hida, K. Disordered ground states in a quantum frustrated spin chain with side chains. *Phys. Rev. B* **2008**, *77*, 134412. [CrossRef]
46. Cheranovskii, V.O.; Slavin, V.V.; Klein, D.J. Quantum phase transitions in frustrated 1D Heisenberg spin systems. *Low Temp. Phys.* **2021**, *47*, 443. [CrossRef]
47. Essler, F.H.L.; Kuzmenko, T. Luttinger liquid coupled to quantum spins: Flow equation approach to the Kondo necklace model. *Phys. Rev. B* **2007**, *76*, 115108. [CrossRef]
48. Hemmatiyan, S.; Movassagh, M.R.; Ghassemi, N.; Kargarian, M.; Rezakhani, A.T.; Langari, A. Quantum phase transitions in the Kondo-necklace model: Perturbative continuous unitary transformation approach. *J. Phys. Condens. Matter* **2015**, *27*, 155601. [CrossRef]
49. Bauer, B.; Carr, L.D.; Evertz, H.G.; Feiguin, A.; Freire, J.; Fuchs, S.; Gamper, L.; Gukelberger, J.; Gull, E.; Guertler, S.; et al. The ALPS project release 2.0: Open source software for strongly correlated systems. *J. Stat. Mech.* **2011**, P05001. [CrossRef]
50. Albuquerque, A.F.; Alet, F.; Corboz, P.; Dayal, P.; Feiguin, A.; Fuchs, S.; Gamper, L.; Gull, E.; Gürtler, S.; Honecker, A.; et al. The ALPS project release 1.3: Open-source software for strongly correlated systems. *J. Magn. Magn. Mater.* **2007**, *310*, 1187. [CrossRef]
51. Lieb, E.; Mattis, D. Ordering Energy Levels of Interacting Spin Systems. *J. Math. Phys.* **1962**, *3*, 749. [CrossRef]

 *materials*

Article

Unraveling the Phase Stability and Physical Property of Modulated Martensite in $Ni_2Mn_{1.5}In_{0.5}$ Alloys by First-Principles Calculations

Xin-Zeng Liang [1], Jing Bai [1,2,3,*], Zi-Qi Guan [1], Yu Zhang [1], Jiang-Long Gu [4], Yu-Dong Zhang [5], Claude Esling [5], Xiang Zhao [1,*] and Liang Zuo [1]

[1] Key Laboratory for Anisotropy and Texture of Materials, Northeastern University, Shenyang 110819, China; 1810167@stu.neu.edu.cn (X.-Z.L.); 13383881381@163.com (Z.-Q.G.); 2110136@stu.neu.edu.cn (Y.Z.); lzuo@mail.neu.edu.cn (L.Z.)

[2] School of Resources and Materials, Northeastern University at Qinhuangdao, Qinhuangdao 066004, China

[3] Hebei Provincial Laboratory for Dielectric and Electrolyte Functional Materials, Qinhuangdao 066004, China

[4] State Key Laboratory of Metastable Materials Science and Technology, Yanshan University, Qinhuangdao 066004, China; gujianglong@ysu.edu.cn

[5] Laboratoire d'Étude des Microstructures et de Mécanique des Matériaux, LEM3 CNRS, UMR 7239, University of Lorraine, 57045 Metz, France; yudong.zhang@univ-lorraine.fr (Y.-D.Z.); claude.esling@univ-lorraine.fr (C.E.)

* Correspondence: baijing@neuq.edu.cn (J.B.); zhaox@mail.neu.edu.cn (X.Z.)

Citation: Liang, X.-Z.; Bai, J.; Guan, Z.-Q.; Zhang, Y.; Gu, J.-L.; Zhang, Y.-D.; Esling, C.; Zhao, X.; Zuo, L. Unraveling the Phase Stability and Physical Property of Modulated Martensite in $Ni_2Mn_{1.5}In_{0.5}$ Alloys by First-Principles Calculations. *Materials* **2022**, *15*, 4032. https://doi.org/10.3390/ma15114032

Academic Editor: Daniela Kovacheva

Received: 23 April 2022
Accepted: 31 May 2022
Published: 6 June 2022

Publisher's Note: MDPI stays neutral with regard to jurisdictional claims in published maps and institutional affiliations.

Abstract: Large magnetic field-induced strains can be achieved in modulated martensite for Ni-Mn-In alloys; however, the metastability of the modulated martensite imposes serious constraints on the ability of these alloys to serve as promising sensor and actuator materials. The phase stability, magnetic properties, and electronic structure of the modulated martensite in the $Ni_2Mn_{1.5}In_{0.5}$ alloy are systematically investigated. Results show that the 6M and 5M martensites are metastable and will eventually transform to the NM martensite with the lowest total energy in the $Ni_2Mn_{1.5}In_{0.5}$ alloy. The physical properties of the incommensurate 7M modulated martensite (7M–IC) and nanotwinned 7M martensite ($7M − (5\bar{2})_2$) are also calculated. The austenite (A) and $7M − (5\bar{2})_2$ phases are ferromagnetic (FM), whereas the 5M, 6M, and NM martensites are ferrimagnetic (FIM), and the FM coexists with the FIM state in the 7M–IC martensite. The calculated electronic structure demonstrates that the splitting of Jahn–Teller effect and the strong Ni–Mn bonding interaction lead to the enhancement of structural stability.

Keywords: Ni–Mn–In; first-principles calculations; modulated martensite; Jahn–Teller effect

1. Introduction

Ferromagnetic shape-memory alloys have attracted great interest due to their properties such as favorable magnetic field-induced strain (MFIS) and magnetocaloric effects (MCEs) [1–5]. Those properties are crucial to the utilization of Ni–Mn-based alloys in applications such as magnetic-driven actuators and solid-state energy-efficient refrigeration. The important factors for achieving large MFIS depend on the type of martensite structure with its c/a ratio around 1.00 [1,6–9]. For example, for modulated martensite with c/a < 1.00, 5.1% and 6% MFIS were obtained in the five layer modulated (5M) martensite [6,7] and 9.5% MFIS in the seven-layer modulated (7M) martensite [1] of the Ni–Mn–Ga alloys. Sozinov et al. [8] achieved a reduction in c/a value in the non-modulated (NM) martensite, from 1.25 [9] to 1.15, by co-doping Co and Cu in the Ni_2MnGa alloy; thus, an MFIS as large as 12% could be obtained.

Austenite (A) can develop modulated (including 5M, six-layer modulated martensite (6M), and 7M) and non-modulated martensite (NM) structures after martensitic transformation in the Ni–Mn-based alloy [10–13]. The observed modulated martensite structures are mainly described by lattice modulation (including commensurate and incommensurate) and nanotwinning (long-period stacking order) [14,15]. The lattice modulation model gives the degree of deviation from equilibrium position for each atom in a periodically amplitude-modulated structure by a modulation equation, e.g., the monoclinic incommensurate model for 7M martensite (7M–IC) [14]; the long-range stacking order model assumes that the atoms in each plane are uniformly sheared, e.g., the $(5\bar{2})_2$ stacking order for 7M martensite $(7M - (5\bar{2})_2)$ [15]. There has been a controversy over the two types of 7M–IC and $7M - (5\bar{2})_2$ martensites due to the complexity of the long-period structure. A large number of experiments on these two types of 7M martensite have been performed [16–25].

The parent phase has an ordered $L2_1$ structure in the Ni–Mn–In alloy and the martensitic transformation shows a non-diffusion type; the modulated martensitic structure in the Ni–Mn–Ga alloy is also extended to the Ni–Mn–In alloy. Righi et al. [23] and Kaufmann et al. [24] stated that the 7M martensite showed a monoclinic 7M–IC model and $7M - (5\bar{2})_2$ nanotwin combination structure for the Ni–Mn–Ga alloy, respectively. Li et al. [25,26] confirmed the monoclinic commensurate structure of the 5M martensite and the monoclinic incommensurate structure of the 7M martensite from the EBSD Kikuchi diffraction patterns. The phase stability and magnetic properties of the commensurate 5M and 7M–IC were subsequently investigated by Xu et al. [27,28] using first-principles calculations based on the experimental results of Li et al.

Liang et al. [29,30] reported that the $Ni_{50}Mn_{37.5}In_{12.5}$ alloy exhibited a 6M martensitic structure at room temperature (RT) by X-ray diffraction (XRD). Krenke et al. [31] determined the crystal structures of the $Ni_{0.5}Mn_{0.5-x}In_x$ ($0.05 \leq x \leq 0.25$) alloys at RT by XRD. When x = 0.05, the alloy presented an NM martensite; for x = 0.10, the crystal structure of the alloy was a monoclinic 7M martensite; and the alloy possessed a monoclinic 5M structure for x = 0.15 and 0.155. Hernando et al. [32] indicated that the $Ni_{50}Mn_{36}In_{14}$ alloy had a 5M martensite structure and the $Mn_{50}Ni_{40}In_{10}$ alloy had a 7M martensitic structure by XRD at 150 K. Yan et al. [33] determined that the 6M martensite possessed a monoclinic incommensurate structure based on neutron diffraction and (3 + 1) D superspace theory in the $Ni_2Mn_{1.44}In_{0.56}$ alloy.

Due to the complexity of the modulated structures, it is difficult to study the phase stability and magnetic properties of different modulated martensitic structures in experiments. Studying the physical properties of modulated martensites by first-principles calculations is a feasible approach. The main purpose of this work is to reveal the phase stability of the $7M - (5\bar{2})_2$ and 7M–IC models existing in experiments by means of the first-principles calculations and to explain the physical nature of the phase stability from the electronic structure. Meanwhile, the austenite (A), 5M, 6M, and NM structures are also taken into account in order to systematically investigate the possible phases experimentally observed in the Ni–Mn–In alloy. This study attempts to comprehend the two experimentally disputed modulation models from a thermodynamic standpoint and provides theoretical support for further research.

2. Computational Methods

The presented calculations were performed with the spin-polarized density-functional theory (DFT) as implemented by the Vienna ab initio Simulation Package (VASP) [34]. The interaction between ions and electrons was described by the projector augmented wave (PAW) method [35], and the exchange–correlation potential was described using the Perdew–Burke–Ernzerhof implementation of a generalized gradient approximation (GGA) [36]. $Ni-3d^84s^2$, $Mn-3d^54s^2$, and $In-4d^{10}5s^25p$ were treated as valence states. The cutoff energy of the plane waves was set to 351 eV. The Brillouin zone was sampled by the Monkhorst–Pack grid [37] with a $10 \times 10 \times 10$ k-point mesh for the A structure, a $7 \times 11 \times 5$ mesh for the 6M structure, an $8 \times 6 \times 4$ mesh for the 5M and 7M structures, and a $7 \times 14 \times 10$ mesh for the NM structure. Due to the difference in the initial lattice constants of the different martensitic structures, the k-point mesh was different based on the Brillouin zone and lattice constants of the austenitic phase. The total energy convergence criterion was set to 10^{-3} eV and the total and atomic forces were set to 0.02 eV/Å for all calculations. For the A and NM structures, 16-atom cells were created, and 40-atom, 24-atom, 56-atom, and 80-atom unit cells were established for the 5M, 6M, $7M - (5\bar{2})_2$, and 7M–IC structures, respectively. The crystal structure model is shown in Figure 1. It should be noted that the modulated martensite models were based on the experimentally resolved structures. Schematic diagrams and detailed atomic Wyckoff positions of the modulated structures involved here are given in Figure S1 and Tables S1–S4 of the supplementary material. The ferromagnetic (FM) and ferrimagnetic (FIM) states were considered for all possible phases; details can be found in Figure S2 of the Supplementary Materials.

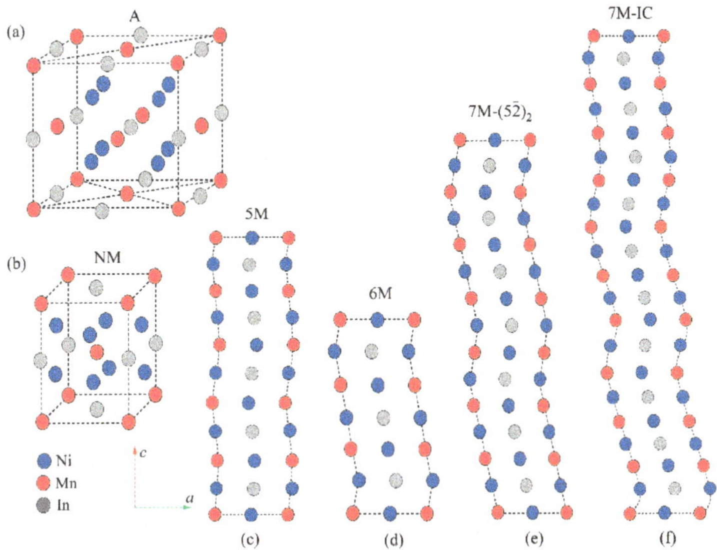

Figure 1. Crystal structures of (**a**) cubic austenite, (**b**) NM martensite, (**c**) 5M martensite, (**d**) 6M martensite, (**e**) $7M - (5\bar{2})_2$ martensite, and (**f**) 7M–IC martensite for Ni_2MnIn alloy.

3. Results and Discussion

3.1. Structural Parameters of Possible Phases

Table 1 shows the equilibrium lattice constants in the FM and FIM states for the possible phases of the $Ni_2Mn_{1.5}In_{0.5}$ alloy. Our calculated result for the A phase in the FM state is 5.95 Å, which is in excellent accordance with the previous theoretical values (5.962 Å [38] and 5.95 Å [39]). Because there is no available experimental evidence for the alloy with the same composition, the XRD results at RT for the $Ni_2Mn_{1.52}In_{0.48}$ and $Ni_2Mn_{1.48}In_{0.52}$ alloys were chosen for comparison with the calculated results for the 7M martensite. The experimental values are relatively close to those obtained from the $7M - (5\bar{2})_2$ structure. However, because the lattice constants are affected by the alloy composition, temperature, and the macroscopic strain field present in the martensite, it remains uncertain which modulated martensite will ultimately be realized in the alloy. Another noteworthy point is that for each structure, the crystal volume (V) of the FM state is larger than that of the FIM state. This is due to the magnetic factor as was noted earlier: the lattice constant of the FM state is greater than that of the non-ferromagnetic state [40].

The optimized lattice constants for the 6M martensite in the FIM state agree with the experimental value measured at T = 300 K using the conventional least-squares approach [29,41]. In particular, the relative error between the theoretically calculated lattice constants for the 6M martensite in the FIM state and those experimentally measured by Wang et al. [41] is only 0.13~0.77%.

Table 1. Theoretical lattice parameters of possible phases of $Ni_2Mn_{1.5}In_{0.5}$ alloy in FM and FIM states in comparison with experimental or other theoretical data.

Structure		Lattice Parameter				
		a (Å)	b (Å)	c (Å)	β (Å)	V
A	FM	5.95, 5.962 [a], 5.95 [b]			90	52.57
	FIM	5.93, 5.94 [a]			90	52.15
5M	FM	4.21	5.91	21.05	90.17	52.42
	FIM	4.41	5.49	21.30	89.03	51.64
6M	FM	4.26	5.82	12.71	91.40	52.51
	FIM	4.41	5.47	12.89	94.07	51.70
	Exp. [c]	4.66	5.40	12.80	95.24	53.49
	Exp. [d]	4.42	5.48	12.99	94.19	
$7M - (5\bar{2})_2$	FM	4.23	5.88	29.55	92.01	52.44
	FIM	4.37	5.55	30.05	95.45	51.78
7M–IC	FM	4.24	5.87	42.28	91.14	52.63
	FIM	4.39	5.52	43.12	94.89	52.02
	Exp. [e]	4.37	5.69	30.21	93.67	
	Exp. [e]	4.35	5.73	30.38	93.24	
NM	FM	4.21		5.95	90	52.64
	FIM	3.87		6.89	90	51.49

[a] Ref. [38], EMTO-CPA. [b] Ref. [39], GGA-PBE [c] Ref. [29], XRD. [d] Ref. [41], XRD. [e] Ref. [42], XRD.

3.2. Phase Stability of Possible Phases

To determine the phase stability of each possible phase in the $Ni_2Mn_{1.5}In_{0.5}$ alloy, the formation energies in the FM and FIM states were calculated and the results are shown in Figure 2a. The formation energy can be calculated as previously reported [43].

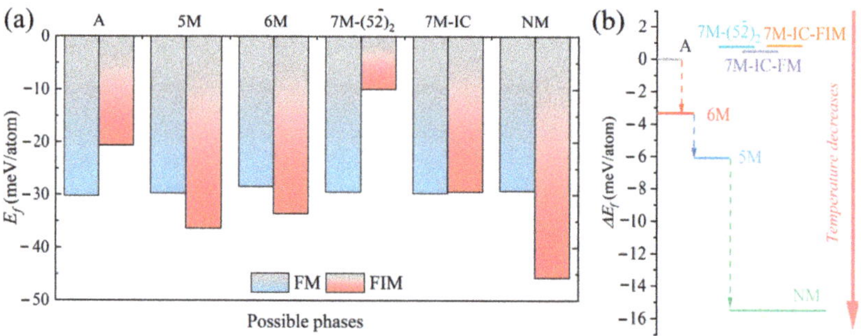

Figure 2. (a) Formation energy of each of the possible phases in FM and FIM states, (b) formation energy difference between each possible martensitic phase and austenite of $Ni_2Mn_{1.5}In_{0.5}$ alloy.

As can be seen from Figure 2a, for both the A and $7M - (5\bar{2})_2$ phases, the formation energy in the FM state is lower than that in the FIM state, indicating that the A and $7M - (5\bar{2})_2$ phases are more likely to possess the FM state; whereas for the 7M–IC martensite, the difference in formation energy between the FM and FIM states is small, only about 0.32 meV/atom, implying that the 7M–IC martensite is strongly susceptible to the co-existence of the FM and FIM states due to incomplete Curie transformation of the martensite. The magnetic ground state of the 7M–IC martensite below is considered to be the FM state for convenience. The formation energy of the FIM state is lower than that of the FM state for the 5M, 6M, and NM martensites, implying that these martensites display the FIM state.

The formation energy difference between austenite and different martensites is also calculated based on the determination of each phase's magnetic ground state; the results are shown in Figure 2b. The formation energies of the two models of the 7M martensite are almost equal, with a difference of only 0.06 meV/atom between the $7M - (5\bar{2})_2$ and 7M–IC in the FM state. This means that the difference in phase stability between these two phases is not significant. It is probable that the macroscopic stress field during the martensitic transformation determines which model of the 7M modulated structure is presented in the experiments. This may be one of the reasons for the controversy between the two models in the experiments. Notice that the reasons for contradictory experimental observations could be also different. For example, it was shown for the 5M martensite in the Ni–Mn–Ga alloys that modulation periodicity changes from commensurate to incommensurate with the decrease in temperature and is accompanied by the refinement of the *a/b* laminate [44–46]. We also found that the formation energy of the 7M martensite is 0.5 meV/atom higher than that of the A phase. This indicates that the 7M martensite is not transformed from the A phase by a thermodynamic driving force. However, the 7M martensite observed in the experiments is likely to be induced by the local stress concentration. For the other martensites, the difference in formation energy is more pronounced. In previous experiments, it was observed that the $Ni_2Mn_{1.5}In_{0.5}$ alloy exhibited 6M martensite at RT [30].

The $Ni_2Mn_{1.5}In_{0.5}$ alloy sample has been previously melted and experimentally characterized by DSC [30], XRD [30], and SEM. The results are shown in Figure 3. As can be seen from Figure 3, the martensitic transformation is observed, and the martensitic transformation temperatures are M_s = 408 K, M_f = 417 K, A_s = 416 K, and A_f = 426 K, respectively. The SEM results show that the alloy presents slatted modulated martensite at room temperature. Furthermore, the martensitic laths have different orientations and different thicknesses in different grains, indicating that different types of martensite may coexist in the $Ni_2Mn_{1.5}In_{0.5}$ alloy at room temperature. The XRD curve of the $Ni_2Mn_{1.5}In_{0.5}$ alloy shows the 6M modulated martensitic structure at room temperature, which is consistent with the calculated results. Combining the results of the first-principles calculations and experiments, it can be seen that the 6M and 5M modulated martensitic structures are metastable; the NM martensite is the most stable structure of the $Ni_2Mn_{1.5}In_{0.5}$ alloy. As confirmed by Dutta et al., the lowest energy structure of martensite is the NM martensite [47].

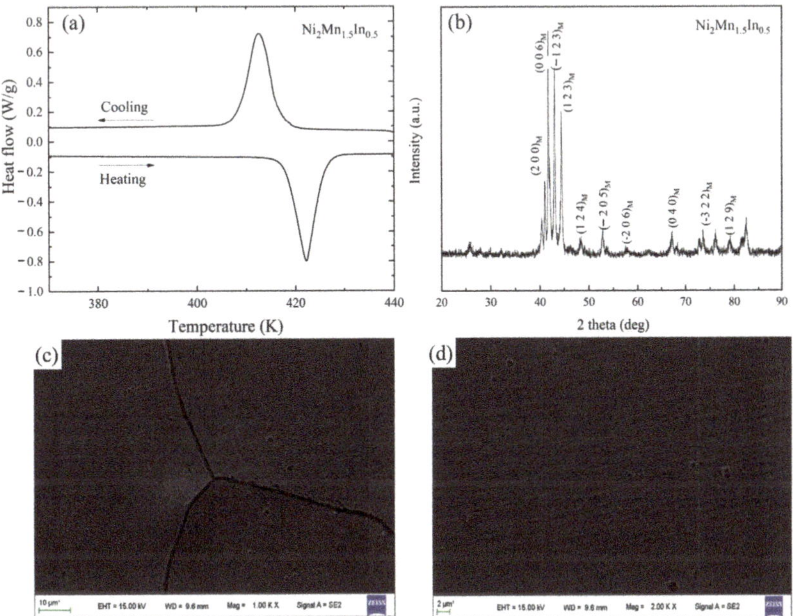

Figure 3. (**a**) DSC curve [30], (**b**) XRD pattern at room temperature [30], (**c**,**d**) microstructures of $Ni_2Mn_{1.5}In_{0.5}$ alloy.

3.3. Total/Atomic Magnetic Moment

The total and atomic magnetic moments of the possible phases in the Ni$_2$Mn$_{1.5}$In$_{0.5}$ alloy are shown in Figure 4a,b, respectively. The total magnetic moments of the A, $7M - (5\bar{2})_2$, and 7M–IC phases in the FM state have little difference. The total magnetic moment of the A phase is about 6.51 μ_B/f.u., which agrees well with the literature values (6.4 μ_B/f.u. [38] and 6.5 μ_B/f.u. [39]). The total magnetic moment decreases abruptly as A transforms to the 6M martensite, indicating that a magnetostructural coupling transformation occurs. The magnetostructural coupling can increase not only the MFIS [48], but also the magnetization difference ΔM [49], thus making such material appealing as a magnetomechanical actuator. The total magnetic moments of the 6M, 5M, and NM are almost the same. The trend in the Ni atomic moment is consistent with the trend in the total magnetic moment, and the magnetic moments of the excess Mn$_{In}$ atoms in the 6M, 5M, and NM phases are all negative, indicating the spin direction of Mn$_{Mn}$ and Mn$_{In}$ present an antiparallel alignment.

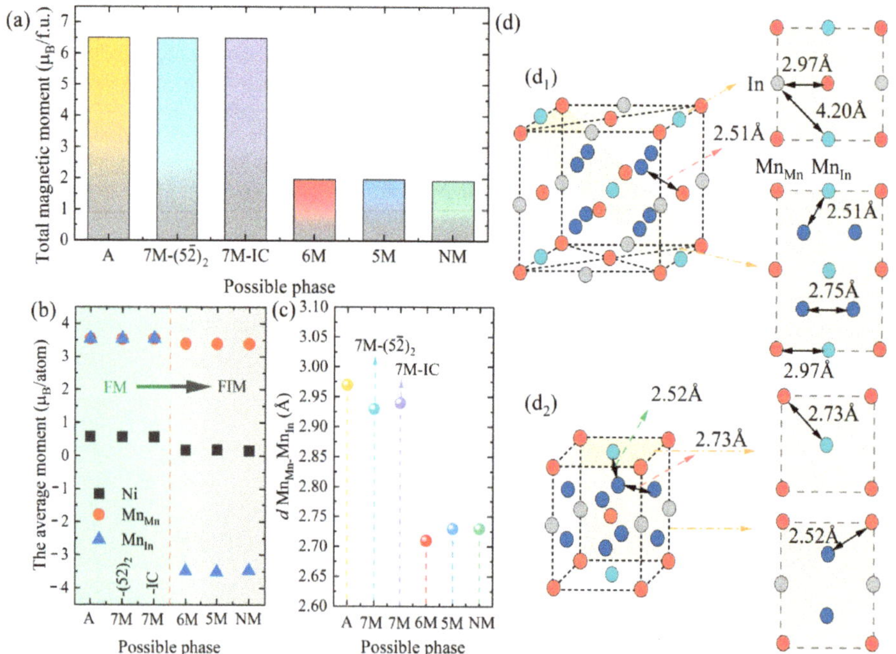

Figure 4. (**a**) Total magnetic moment, (**b**) average of Ni, Mn$_{Mn}$, and Mn$_{In}$ moments, (**c**) nearest-neighbor atomic distance between Mn$_{Mn}$ and Mn$_{In}$ atoms of Ni$_2$Mn$_{1.5}$In$_{0.5}$ alloy, and (**d**) schematic diagram of the nearest atomic distance for (**d$_1$**) A phase and (**d$_2$**) NM phase.

To investigate the underlying reason for the change in the magnetic ground state of each phase, we calculated the nearest-neighbor atomic distances in the A and NM phases, as shown in Figure 4d. It can be seen that the atomic distances of Ni-Mn$_{Mn}$, Ni-Mn$_{In}$, and Ni-Ni remain almost constant during the A→NM transformation (2.51, 2.51, 2.75 Å for the A phase and 2.52, 2.52, 2.73 Å for the NM phase, respectively); whereas the Mn$_{Mn}$-Mn$_{In}$ atomic distance (d (Mn$_{Mn}$-Mn$_{In}$)) decreases from 2.97 Å to 2.73 Å. The d (Mn$_{Mn}$-Mn$_{In}$) for the possible phases are summarized in Figure 4c. This indicates that the shortening of d (Mn$_{Mn}$-Mn$_{In}$) leads to enhanced interaction between the Mn$_{Mn}$ and Mn$_{In}$ atoms, resulting in a magnetostructural coupling transformation.

3.4. Electronic Structure

To understand the physical nature of the relative stability of each martensite, the total densities of states (DOS) and the differential charge densities of the possible phases are shown in Figure 5. The relative stability of the different structures can be understood not only by the features near the Fermi level (E_F) [50–52], but can also be influenced by the bonding ability between Ni and Mn [27,53].

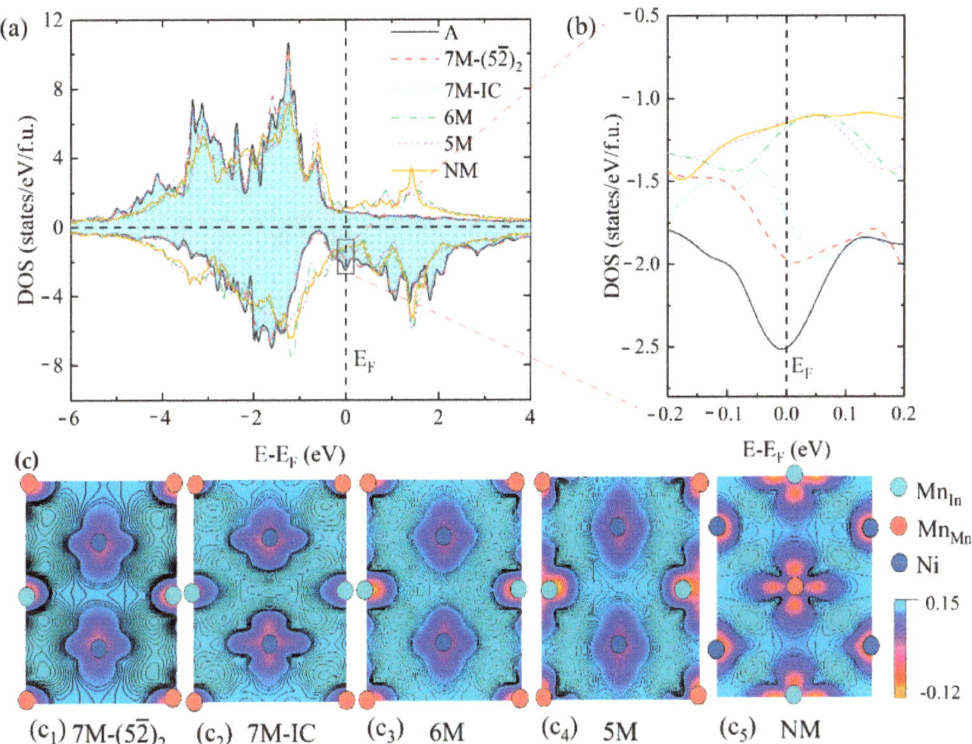

Figure 5. (**a**) Total density of states of A, 5M, 6M, $7M-(5\bar{2})_2$, 7M–IC, and NM phases of $Ni_2Mn_{1.5}In_{0.5}$ alloy, (**b**) enlarged spin-down density of states near the E_F, and (**c**) differential charge densities of different martensitic structures in plane with excess Mn_{In} atoms.

It can be seen from Figure 5a that the main change in the density of states is concentrated near the E_F. The total DOS near the E_F is enlarged, as shown in Figure 5b. The peak of the A phase is located exactly at the E_F, and the spin-down density of states at the E_F is the largest. This indicates that the A phase is extremely unstable. As the symmetry decreases, the states at the E_F are redistributed due to martensitic transformation. The $7M-(5\bar{2})_2$ and 7M–IC martensites also have peaks near the E_F, but their numbers of states are lower than that of the A phase. However, for the 6M, 5M, and NM martensites, the pseudopotential valleys appear at the E_F. This suggests that a Jahn–Teller effect [54–56] occurs in the alloy as the martensitic transformation takes place, which stabilizes the 6M, 5M, and NM martensites.

As can be seen in Figure 5c, the bonding behavior not only exists between Ni-Mn_{Mn}, but also for Ni-Mn_{In} for all the martensitic structures. The bonding ability of Ni-Mn_{Mn}(Mn_{In}) in the modulated martensite is not significantly different. However, the bonding ability between Ni-Mn_{Mn}(Mn_{In}) in the NM martensite is stronger than that in the modulated martensite. Therefore, the bonding ability between Ni and Mn also plays an important role in phase stability.

4. Conclusions

Based on first-principles calculations, a comprehensive study of the structural and electronic properties of the $Ni_2Mn_{1.5}In_{0.5}$ alloy was carried out. The phase stability and magnetic properties of the experimentally observed 5M, 6M, 7M–IC, $7M - (5\bar{2})_2$, and NM martensitic structures were investigated. The calculated equilibrium lattice constants are in good agreement with those determined by experiments and theoretical calculations. For the 7M martensite, the formation energies of the two models are very close. The A and $7M - (5\bar{2})_2$ phases possess FM states; the FM and FIM states co-exist in the 7M–IC martensite; and the 5M, 6M, and NM martensites prefer to exhibit the FIM states. The alloy undergoes a magnetostructural coupling transformation, which is attributed to the shortening of the Mn_{Mn}-Mn_{In} atomic distance. The phase stability is dependent on the Jahn–Teller effect and the bonding behavior between Ni and Mn.

Supplementary Materials: The following supporting information can be downloaded at: https://www.mdpi.com/article/10.3390/ma15114032/s1, see Supplementary Materials for the schematic diagrams and detailed atomic Wyckoff positions of the modulated structures and the setting of magnetic configurations. References [20,23,33,57] is cited in the supplementary materials.

Author Contributions: Conceptualization, X.Z. and L.Z.; funding acquisition, J.B.; investigation, X.-Z.L.; supervision, X.Z. and L.Z.; validation, J.-L.G. and Y.Z.; visualization, J.B. and Z.-Q.G., Y.-D.Z. and C.E.; writing—original draft, X.-Z.L.; writing—review & editing, X.-Z.L., Z.-Q.G., Y.-D.Z., J.-L.G., Y.Z. and C.E. All authors have read and agreed to the published version of the manuscript.

Funding: This work is supported by the National Natural Science Foundation of China (Grant No. 51771044), the Fundamental Research Funds for the Central Universities (No. N2223025), the State Key Lab of Advanced Metals and Materials (No. 2022-Z02), and Programme of Introducing Talents of Discipline Innovation to Universities 2.0 (the 111 Project of China 2.0, No. BP0719037). This work was carried out at Shanxi Supercomputing Center of China, and the calculations were performed on TianHe-2.

Institutional Review Board Statement: Not applicable.

Informed Consent Statement: Not applicable.

Data Availability Statement: The data that support the findings of this study are available from the corresponding author upon reasonable request.

Conflicts of Interest: The authors declare no conflict of interest.

References

1. Sozinov, A.; Likhachev, A.; Lanska, N.; Ullakko, K. Giant magnetic-field-induced strain in NiMnGa seven-layered martensitic phase. *Appl. Phys. Lett.* **2002**, *80*, 1746–1748. [CrossRef]
2. Chmielus, M.; Zhang, X.; Witherspoon, C.; Dunand, D.; Müllner, P. Giant magnetic-field-induced strains in polycrystalline Ni-Mn-Ga foams. *Nat. Mater.* **2009**, *8*, 863–866. [CrossRef]
3. Ullakko, K.; Huang, J.; Kantner, C.; Ohandley, R.; Kokorin, V. Large magnetic-field-induced strains in Ni_2MnGa single crystals. *Appl. Phys. Lett.* **1996**, *69*, 1966–1968. [CrossRef]
4. Yang, J.; Li, Z.; Yang, B.; Yan, H.; Cong, D.; Zhao, X.; Zuo, L. Effects of Co and Si co-doping on magnetostructural transformation and magnetocaloric effect in Ni-Mn-Sn based alloys. *J. Alloys Comp.* **2022**, *892*, 162190. [CrossRef]
5. Pfeuffer, L.; Lemke, J.; Shayanfar, N.; Riegg, S.; Koch, D.; Taubel, A.; Scheibel, F.; Kani, N.A.; Adabifiroozjaei, E.; Molina-Luna, L.; et al. Microstructure engineering of metamagnetic Ni-Mn-based Heusler compounds by Fe-doping: A roadmap towards excellent cyclic stability combined with large elastocaloric and magnetocaloric effects. *Acta Mater.* **2021**, *221*, 117390. [CrossRef]
6. Heczko, O.; Sozinov, A.; Ullakko, K. Giant field-induced reversible strain in magnetic shape memory NiMnGa alloy. *IEEE Trans. Magn.* **2000**, *36*, 3266–3268. [CrossRef]
7. Murray, S.J.; Marioni, M.; Allen, S.M.; O'Handley, R.C.; Lograsso, T.A. 6% magnetic-field-induced strain by twin-boundary motion in ferromagnetic Ni-Mn-Ga. *Appl. Phys. Lett.* **2000**, *77*, 886–888. [CrossRef]
8. Sozinov, A.; Lanska, N.; Soroka, A.; Zou, W. 12% magnetic field-induced strain in Ni-Mn-Ga-based non-modulated martensite. *Appl. Phys. Lett.* **2013**, *102*, 021902. [CrossRef]

9. Yan, H.L.; Liu, H.X.; Zhao, Y.; Jia, N.; Bai, J.; Yang, B.; Li, Z.B.; Zhang, Y.D.; Esling, C.; Zhao, X.; et al. Impact of B alloying on ductility and phase transition in the Ni–Mn-based magnetic shape memory alloys: Insights from first-principles calculation. *J. Mater. Sci. Technol.* **2021**, *74*, 27–34. [CrossRef]

10. Yan, H.L.; Huang, X.M.; Esling, C. Recent Progress in Crystallographic Characterization, Magnetoresponsive and Elastocaloric Effects of Ni-Mn-In-Based Heusler Alloys—A Review. *Front. Mater.* **2022**, *85*, 812984. [CrossRef]

11. Akır, A.C.; Righi, L.; Albertini, F.; Acet, M.; Farle, M.; Akturk, S. Extended investigation of intermartensitic transitions in Ni-Mn-Ga magnetic shape memory alloys: A detailed phase diagram determination. *J. Appl. Phys.* **2013**, *114*, 183912.

12. Singh, S.; Petricek, V.; Rajput, P.; Hill, A.H.; Suard, E.; Barman, S.R.; Pandey, D. High-resolution synchrotron x-ray powder diffraction study of the incommensurate modulation in the martensite phase of Ni_2MnGa: Evidence for nearly 7M modulation and phason broadening. *Phys. Rev. B* **2014**, *90*, 014109. [CrossRef]

13. Wang, H.; Li, D.; Zhang, G.; Li, Z.; Yang, B.; Yan, H.; Cong, D.; Esling, C.; Zhao, X.; Zuo, L. Highly sensitive elastocaloric response in a directionally solidified $Ni_{50}Mn_{33}In_{15.5}Cu_{1.5}$ alloy with strong <001> A preferred orientation. *Intermetallics* **2022**, *140*, 107379. [CrossRef]

14. Martynov, V.V.; Kokorin, V.V. The crystal structure of thermally-and stress-induced martensites in Ni_2MnGa single crystals. *J. Phys. III* **1992**, *2*, 739–749. [CrossRef]

15. Otsuka, K.; Ohba, T.; Tokonami, M.; Wayman, C.M. New description of long period stacking order structures of martensites in β phase alloys. *Scr. Metall. Mater.* **1993**, *29*, 1359–1364. [CrossRef]

16. Brown, P.J.; Crangle, J.; Kanomata, T.; Matsmuoto, M.; Neumann, K.U.; Ouladdiaf, B.; Ziebeck, K.R.A. The crystal structure and phase transitions of the magnetic shape memory compound Ni_2MnGa. *J. Phys. Condens. Matter* **2002**, *14*, 10159–10171. [CrossRef]

17. Mariager, S.; Huber, T.; Ingold, G. The incommensurate modulations of stoichiometric Ni_2MnGa. *Acta Mater.* **2014**, *66*, 192–198. [CrossRef]

18. Righi, L.; Albertini, F.; Calestani, G.; Pareti, L.; Paoluzi, A.; Ritter, C.; Algarabel, P.A.; Morellon, L.; Ricardo Ibarra, M. Incommensurate modulated structure of the ferromagnetic shape-memory Ni_2MnGa martensite. *J. Solid State Chem.* **2006**, *179*, 3525–3533. [CrossRef]

19. Righi, L.; Albertini, F.; Pareti, L.; Paoluzi, A.; Calestani, G. Commensurate and incommensurate "5M" modulated crystal structures in Ni-Mn-Ga martensitic phases. *Acta Mater.* **2007**, *55*, 5237–5245. [CrossRef]

20. Glavatskyy, I.; Glavatska, N.; Urubkov, I.; Hoffman, J.U.; Bourdarot, F. Crystal and magnetic structure temperature evolution in Ni-Mn-Ga magnetic shape memory martensite. *Mater. Sci. Eng. A* **2008**, *481-482*, 298–301. [CrossRef]

21. Gruner, M.E.; Fahler, S.; Entel, P. Magnetoelastic coupling and the formation of adaptive martensite in magnetic shape memory alloys. *Phys. Status Solodi B* **2014**, *251*, 2067–2079. [CrossRef]

22. Zayak, A.T.; Entel, P.; Enkovaara, J.; Ayuela, A.; Nieminen, R.M. First-principles investigations of homogeneous lattice-distortive strain and shuffles in Ni_2MnGa. *J. Phys. Condens. Matter* **2003**, *15*, 159–164. [CrossRef]

23. Righi, L.; Albertini, F.; Villa, E.; Paoluzi, A.; Calestani, G.; Chernenko, V.; Besseghini, S.; Ritter, C.; Passaretti, F. Crystal structure of 7M modulated Ni-Mn-Ga martensitic phase. *Acta Mater.* **2008**, *56*, 4529–4535. [CrossRef]

24. Kaufmann, S.; Rößler, U.K.; Heczko, O.; Wuttig, M.; Buschbeck, J.; Schultz, L.; Fähler, S. Adaptive Modulations of Martensites. *Phys. Rev. Lett.* **2010**, *104*, 145702. [CrossRef]

25. Li, Z.B.; Zhang, Y.D.; Esling, C.; Zhao, X.; Zuo, L. Twin relationships of 5M modulated martensite in Ni-Mn-Ga alloy. *Acta Mater.* **2011**, *59*, 3390–3397. [CrossRef]

26. Li, Z.B.; Zhang, Y.D.; Esling, C.; Zhao, X.; Zuo, L. Determination of the orientation relationship between austenite and incommensurate 7M modulated martensite in Ni-Mn-Ga alloys. *Acta Mater.* **2011**, *59*, 2762–2772. [CrossRef]

27. Xu, N.; Raulot, J.M.; Li, Z.B.; Zhang, Y.D.; Bai, J.; Peng, W.; Meng, X.Y.; Zhao, X.; Zuo, L.; Esling, C. Composition dependent phase stability of Ni-Mn-Ga alloys studied by ab initio calculations. *J. Alloys Comp.* **2014**, *614*, 126–130. [CrossRef]

28. Bai, J.; Wang, J.L.; Shi, S.F.; Raulot, J.M.; Zhang, Y.D.; Esling, C.; Zhao, X.; Zuo, L. Complete martensitic transformation sequence and magnetic properties of non-stoichiometric $Ni_2Mn_{1.2}Ga_{0.8}$ alloy by first-principles calculations. *J. Magn. Magn. Mater.* **2019**, *473*, 360–364. [CrossRef]

29. Liang, X.Z.; Bai, J.; Gu, J.L.; Wang, J.L.; Yan, H.L.; Zhang, Y.D.; Esling, C.; Zhao, X.; Zuo, L. Ab initio-based investigation of phase transition path and magnetism of Ni-Mn-In alloys with excess Ni or Mn. *Acta Mater.* **2020**, *195*, 109–122. [CrossRef]

30. Liang, X.Z.; Bai, J.; Gu, J.L.; Yan, H.L.; Zhang, Y.D.; Esling, C.; Zhao, X.; Zuo, L. Probing martensitic transformation, kinetics, elastic and magnetic properties of $Ni_{2-x}Mn_{1.5}In_{0.5}Co_x$ alloys. *J. Mater. Sci. Technol.* **2020**, *44*, 31–41. [CrossRef]

31. Krenke, T.; Acet, M.; Wassermann, E.F.; Moya, X.; Manosa, L.; Planes, A. Ferromagnetism in the austenitic and martensitic states of Ni-Mn-In alloys. *Phys. Rev. B* **2006**, *73*, 174413. [CrossRef]

32. Hernando, B.; Llamazares, J.L.S.; Santos, J.D.; Sanchez, M.L.; Escoda, L.; Sunol, J.J.; Varga, R.; Garcia, C.; Gonzalez, J. Grain oriented NiMnSn and NiMnIn Heusler alloys ribbons produced by melt spinning: Martensitic transformation and magnetic properties. *J. Magn. Magn. Mater.* **2009**, *321*, 763–768. [CrossRef]

33. Yan, H.L.; Zhang, Y.D.; Xu, N.; Senyshyn, A.; Brokmeier, H.G.; Esling, C.; Zhao, X.; Zuo, L. Crystal structure determination of incommensurate modulated martensite in Ni-Mn-In Heusler alloys. *Acta Mater.* **2015**, *88*, 375–388. [CrossRef]

34. Kresse, G.; Joubert, D. From ultrasoft pseudopotentials to the projector augmented-wave method. *Phys. Rev. B* **1999**, *59*, 1758–1775. [CrossRef]

35. Blochl, P.E. Projector augmented-wave method. *Phys. Rev. B Condens. Matter* **1994**, *50*, 17953–17979. [CrossRef]

36. Perdew, J.P.; Burke, K.; Ernzerhof, M. Generalized gradient approximation made simple. *Phys. Rev. Lett.* **1996**, *77*, 3865–3868. [CrossRef]
37. Monkhorst, H.J.; Pack, J.D. Special points for Brillouin-zone integrations. *Phys. Rev. B* **1976**, *13*, 5188–5192. [CrossRef]
38. Li, C.M.; Luo, H.B.; Hu, Q.M.; Yang, R.; Johansson, B.; Vitos, L. Role of magnetic and atomic ordering in the martensitic transformation of Ni-Mn-In from a first-principles study. *Phys. Rev. B* **2012**, *86*, 214205. [CrossRef]
39. He, W.Q.; Huang, H.B.; Liu, Z.H.; Ma, X.Q. First-principles investigation of magnetic properties and metamagnetic transition of NiCoMnZ (Z = In, Sn, Sb) Heusler alloys. *Intermetallics* **2017**, *90*, 140–146. [CrossRef]
40. Li, C.M.; Luo, H.B.; Hu, Q.M.; Yang, R.; Johansson, B.; Vitos, L. First-principles investigation of the composition dependent properties of $Ni_{2+x}Mn_{1-x}Ga$ shape-memory alloys. *Phys. Rev. B* **2010**, *82*, 024201. [CrossRef]
41. Wang, C.H. Crystal Structure, Martensitic Transformation and Microstructure in Ni-Mn-In Meta-Magnetic Shape Memory Alloys. Master's Thesis, Northeastern University, Shenyang, China, 2015.
42. Feng, Y. Martensitic Transformation Behavior and Structure and Properties of Ni-Mn-In Based Alloys. Ph.D. Thesis, Harbin Institute of Technology, Harbin, China, 2009.
43. Xu, N.; Raulot, J.M.; Li, Z.B.; Bai, J.; Zhang, Y.D.; Zhao, X.; Zuo, L.; Esling, C. Oscillation of the magnetic moment in modulated martensites in Ni_2MnGa studied by ab initio calculations. *Appl. Phys. Lett.* **2012**, *100*, 084106. [CrossRef]
44. Straka, L.; Drahokoupil, J.; Vertat, P.; Kopecek, J.; Zeleny, M.; Seiner, H.; Heczko, O. Orthorhombic intermediate phase originating from {110} nanotwinning in $Ni_{50.0}Mn_{28.7}Ga_{21.3}$ modulated martensite. *Acta. Mater.* **2017**, *132*, 335–344. [CrossRef]
45. Straka, L.; Drahokoupil, J.; Vertat, P.; Zeleny, M.; Kopecek, J.; Sozinov, A.; Heczko, O. Low temperature a/b nanotwins in $Ni_{50}Mn_{25+x}Ga_{25-x}$ Heusler alloys. *Sci. Rep.* **2018**, *8*, 11943. [CrossRef] [PubMed]
46. Vertat, P.; Seiner, H.; Straka, L.; Klicpera, M.; Sozinov, A.; Fabelo, O.; Heczko, O. Hysteretic structural changes within five-layered modulated 10M martensite o fNi–Mn–Ga (–Fe). *J. Phys. Condens. Matter* **2021**, *33*, 265404. [CrossRef]
47. Dutta, B.; Cakir, A.; Giacobbe, C.; Al-Zubi, A.; Hickel, T.; Acet, M.; Neugebauer, J. Ab initio prediction of martensitic and intermartensitic phase boundaries in Ni-Mn-Ga. *Phys. Rev. Lett.* **2016**, *116*, 025503. [CrossRef]
48. Marcos, J.; Planes, A.; Mañosa, L.; Casanova, F.; Batlle, X.; Labarta, A.; Martínez, B. Magnetic field induced entropy change and magnetoelasticity in Ni-Mn-Ga alloys. *Phys. Rev. B* **2002**, *66*, 224413. [CrossRef]
49. Wu, Z.; Liu, Z.; Yang, H.; Liu, Y.; Wu, G. Effect of Co addition on martensitic phase transformation and magnetic properties of $Mn_{50}Ni_{40-x}In_{10}Co_x$ polycrystalline alloys. *Intermetallics* **2011**, *19*, 1839–1848. [CrossRef]
50. Kundu, A.; Gruner, M.E.; Siewert, M.; Hucht, A.; Entel, P.; Ghosh, S. Interplay of phase sequence and electronic structure in the modulated martensites of Mn_2NiGa from first-principles calculations. *Phys. Rev. B* **2017**, *96*, 064107. [CrossRef]
51. Ghosh, S.; Ghosh, S. Role of composition, site ordering, and magnetic structure for the structural stability of off-stoichiometric Ni_2MnSb alloys with excess Ni and Mn. *Phys. Rev. B* **2019**, *99*, 064112. [CrossRef]
52. Kundu, A.; Ghosh, S. Site occupancy, composition and magnetic structure dependencies of martensitic transformation in $Mn_2Ni_{1+x}Sn_{1-x}$. *J. Phys. Condens. Matter* **2018**, *30*, 015401. [CrossRef]
53. Liang, X.Z.; Jiang, X.J.; Gu, J.L.; Bai, J.; Guan, Z.Z.; Li, Z.Z.; Yan, H.L.; Zhang, Y.D.; Esling, C.; Zhao, X.; et al. 5M and 7M martensitic stability and associated physical properties in $Ni_{50}Mn_{35}In_{15}$ alloy: First-principles calculations and experimental verification. *Scr. Mater.* **2021**, *204*, 114140. [CrossRef]
54. Barman, S.R.; Banik, S.; Chakrabarti, A. Structural and electronic properties of Ni_2MnGa. *Phys. Rev. B* **2005**, *72*, 184410. [CrossRef]
55. Liang, Z.; Li, Q.; Sun, K.; Luo, H.J. Martensitic transition and magnetic structure in Zn-doped Heusler alloy Mn_2NiGa: A theoretical approach. *Phys. Chem. Solids* **2019**, *134*, 127–132. [CrossRef]
56. Zelený, M.; Sozinov, A.; Straka, L.; Björkman, T.; Nieminen, R.M. First-principles study of Co- and Cu-doped along the tetragonal deformation path. *Phys. Rev. B* **2014**, *89*, 184103. [CrossRef]
57. Li, Z.B. Study on Crystallographic Features of Ni-Mn-Ga Ferromagnetic Shape Memory Alloys. Ph.D. Thesis, Paul-Verlaine University of Metz, Metz, France, Northeastern University, Shenyang, China, 2011.

Article

Microwave-Assisted Solvothermal Synthesis of Nanocrystallite-Derived Magnetite Spheres

Greta Zambzickaite [1], Martynas Talaikis [1], Jorunas Dobilas [2], Voitech Stankevic [2], Audrius Drabavicius [3], Gediminas Niaura [1,*] and Lina Mikoliunaite [1,4,*]

1 Department of Organic Chemistry, Center for Physical Sciences and Technology (FTMC), Sauletekio al. 3, 10257 Vilnius, Lithuania; greta.zambzickaite@ftmc.lt (G.Z.); martynas.talaikis@ftmc.lt (M.T.)
2 Department of Functional Materials and Electronics, Center for Physical Sciences and Technology (FTMC), Sauletekio al. 3, 10257 Vilnius, Lithuania; jorunas.dobilas@ftmc.lt (J.D.); voitech.stankevic@ftmc.lt (V.S.)
3 Department of Characterization of Materials Structure, Center for Physical Sciences and Technology (FTMC), Sauletekio al. 3, 10257 Vilnius, Lithuania; audrius.drabavicius@ftmc.lt
4 Department of Physical Chemistry, Faculty of Chemistry and Geosciences, Vilnius University, Naugarduko st. 24, 03225 Vilnius, Lithuania
* Correspondence: gediminas.niaura@ftmc.lt (G.N.); lina.mikoliunaite@ftmc.lt (L.M.)

Abstract: The synthesis of magnetic particles triggers the interest of many scientists due to their relevant properties and wide range of applications in the catalysis, nanomedicine, biosensing and magnetic separation fields. A fast synthesis of iron oxide magnetic particles using an eco-friendly and facile microwave-assisted solvothermal method is presented in this study. Submicron Fe_3O_4 spheres were prepared using $FeCl_3$ as an iron source, ethylene glycol as a solvent and reductor and sodium acetate as a precipitating and nucleating agent. The influence of the presence of polyethylene glycol as an additional reductor and heat absorbent was also evaluated. We reduce the synthesis time to 1 min by increasing the reaction temperature using the microwave-assisted solvothermal synthesis method under pressure or by adding PEG at lower temperatures. The obtained magnetite spheres are 200–300 nm in size and are composed of 10–30 nm sized crystallites. The synthesized particles were investigated using the XRD, TGA, pulsed-field magnetometry, Raman and FTIR methods. It was determined that adding PEG results in spheres with mixed magnetite and maghemite compositions, and the synthesis time increases the size of the crystallites. The presented results provide insights into the microwave-assisted solvothermal synthesis method and ensure a fast route to obtaining spherical magnetic particles composed of different sized nanocrystallites.

Keywords: magnetite; Fe_3O_4; microwave-assisted solvothermal method

Citation: Zambzickaite, G.; Talaikis, M.; Dobilas, J.; Stankevic, V.; Drabavicius, A.; Niaura, G.; Mikoliunaite, L. Microwave-Assisted Solvothermal Synthesis of Nanocrystallite-Derived Magnetite Spheres. *Materials* **2022**, *15*, 4008. https://doi.org/10.3390/ma15114008

Academic Editors: Haiou Wang and Dexin Yang

Received: 18 May 2022
Accepted: 2 June 2022
Published: 5 June 2022

Publisher's Note: MDPI stays neutral with regard to jurisdictional claims in published maps and institutional affiliations.

1. Introduction

Magnetic particles are highly desirable in many scientific fields, especially biomedical fields [1,2], starting from MRI contrast agents [3,4], drug delivery systems [5,6], magnetic separators [7] and hyperthermia agents [8] and followed by the environmental [9], catalysis [10] and biosensing [11,12] fields. These nanoparticles capped with a plasmonic silver or gold layer could also be applied in surface-enhanced Raman spectroscopy due to signal enhancement for two reasons: the concentration of the sample using magnet and surface plasmons [13–15]. Magnetite particles are suitable for such applications due to their stability, biocompatibility, uncomplicated synthesis, low price and great magnetic response. However, the development of synthesis methods is still extremely important and not completely clear in obtaining a tailor-made product.

Many synthesis routs can be applied to obtain magnetic particles. They can be divided into physical, biological and chemical methods [2]. The physical methods are usually called top-down methods, where bulk material is shredded to smaller pieces by ball milling, laser evaporation or any other physical method. The biological synthesis routes employ biological objects that, in the presence of Fe ions in a solution, could produce nanoparticles within their structure or in the solution. Chemical synthesis, otherwise known as the bottom-up method, is the most widely applied in the synthesis of magnetic structures. This route could be further divided according to the synthesis procedure: hydrothermal [16], solvothermal [12,17], thermal decomposition [18], microwave-assisted [19,20], sol-gel [21,22], coprecipitation [23] and others [4]. In addition to this, the combination of two methods could be used: microwave-assisted solvothermal [24–29] or hydrothermal [20,30] methods or microwave assisted thermal decomposition methods [31].

Microwave synthesis is now gaining popularity due to its simplicity and reduced processing time. Microwave heating, which is mainly described as heat obtained from transformed electromagnetic energy, is used to accelerate the synthesis procedure instead of conventional heating. If the latter is used, a sample is heated starting from the sides of the reaction vessel due to the conduction and convection processes; thus, the temperature gradient appears, which causes non-uniform particle nucleation and growing processes. Microwave heating does not pose such a problem, since the energy is transferred through the sample volume (or microwave absorbing material) instantly and the heating is homogeneous [20]. During microwave heating, two effects can be distinguished: thermal and non-thermal. The first one is considered as fast and homogeneous and thus as an effective heating method of the sample volume, resulting in nucleation and growth. This gives uniform and high crystallinity results. During the non-thermal effect, hot spots and hot surfaces are created when heating solid surfaces at the solid–liquid interface. This process also supports the reduction of precursors, nucleation and the formation of particles [19]. In the microwave synthesis route, both organic and inorganic media could be used, and the synthesis time could be reduced to minutes instead of hours or days. Therefore, it is an energy-saving method as well. The microwave-assisted solvothermal method provides the possibility to reduce the synthesis time efficiently [26,27]. In addition, the modification of nanoparticles by coating with 3-(trimethoxysilyl)-1-propanethiol (TMSPT) and the subsequent modification by coating with 2-amino-5-mercapto-1,3,4-thiadiazole (AMP) to increase the stability of the nanoparticles are possible through this technique [26]. Thus, the whole process, including the synthesis of the nanoparticles, the coating with TMSPT and the modification with AMP, was accomplished during a short period of time (30 min) [26]. Furthermore, the semicrystalline Fe_3O_4 nanoparticles with an average diameter of 15 nm generated by the microwave-assisted solvothermal process were demonstrated to be active in the photocatalytic degradation of azo dye methyl orange and tetracycline under visible light radiation [27].

The medium of the synthesis varies from the most common, water, to organic solvent (e.g., glycol). All of these conditions severely affect the resulting material properties. Particles of various shapes, sizes, crystal structures and magnetization extents could be obtained. The shapes vary from cubic to spherical or rodlike [32]. Sizes from a few nanometers to several hundred nanometers could be achieved [33,34]. A few iron oxide crystal structures are known: the hematite (α-Fe_2O_3)—rhombohedral or hexagonal; the maghemite (γ-Fe_2O_3)—cubic or tetrahedral; and the magnetite (Fe_3O_4 or $FeO \cdot Fe_2O_3$)—cubic [35]. They are most commonly found in nature, soil and rocks from volcanic eruptions, as well as from air pollution (emissions from traffic and industries) and from being synthesized in the laboratory.

Another highly important compound is the reducing agent (NaOH, ammonia and H_2O_2 are the most popular in aqueous solutions or glycol in nonaqueous synthesis) [6,15,23,33,36,37]. Finally, a stabilizing/capping agent may also be used. This could be some surfactant (for electrostatic stabilization such as sodium oleate, dodecylamine and sodium carboxymethyl cellulose) or polymer (for steric and, in some cases, electrostatic stabilization—for example, the chemical polymers polyethylene glycol, poly(vinyl alcohol), poly(lactic-co-glycolic acid), poly(vinyl-pyrrolidone), poly(ethylene-co-vinyl acetate) and other polymers [38], or natural polymer systems including gelatin, dextran and chitosan [39].

Polyethylene glycol is a biocompatible, hydrophilic, water-soluble organic polymer. Due to its high polarizability, it is an excellent microwave-absorbing agent that ensures a high heating rate and a significantly shorter reaction time. Harraz and colleagues synthesized single-phase magnetic nanowires using PEG in aqueous media and noticed that the amount of PEG in the reaction solution affects the size and crystallinity of the obtained results [40]. The porosity of the obtained nanoparticles during the microwave-assisted solvothermal synthesis was evaluated by Juang et al. [24].

In this work, organic magnetite synthesis in a microwave reactor is investigated. A combination of two methods, solvothermal and microwave, was used, resulting in microwave-assisted solvothermal synthesis. The advantages of such synthesis route include the saving of time and energy. In addition to this, benignancy to the environment is also an advantage, since the process is carried out in a closed system and a strong acid/base was not used in the initial solutions. It is known that microwave radiation is a great source of energy that offers a clean and effective form of heating [24]. The reaction was conducted in ethylene glycol, which acted as a reducing agent, as well in the presence of acetate, which acted as a nucleating/capping agent [25]. Additionally, the impact of polyethylene glycol was evaluated. During the synthesis, spherical magnetic particles with different morphologies were obtained. The microwave synthesis conditions such as the temperature and time were investigated. The structural, morphological and magnetic characterizations of the obtained particles are presented. The novelty of our work consists in the development of a simple microwave-assisted solvothermal method for the synthesis of spherical magnetic Fe_3O_4 particles consisting of nanocrystallites that are $10-30$ nm in size and a systematic investigation on the temperature, time and availability of PEG, resulting in a successful reaction. The obtained results could provide more insight into the microwave-assisted solvothermal synthesis, providing a very fast route to obtaining spherical magnetic submicron sized particles composed of different sized nanocrystallites with a magnetite or magnetite/maghemite mix structure and helping to analyze their physical and chemical properties.

2. Materials and Methods

2.1. Materials

Polyethylene glycol (PEG, MW 20000) and ethylene glycol (EG) were purchased from Carl Roth GmbH (Karlsruhe, Germany); sodium acetate ($NaCH_3COO$; NaOAc) was obtained from Alfa Aesar (Haverhill, MA, USA); and iron(III) chloride ($FeCl_3$) was obtained from Sigma Aldrich (St. Louis, MO, USA). All of the chemicals were of analytical grade and were used as obtained.

2.2. Synthesis of Magnetic Fe_3O_4 Particles

The synthesis of magnetite (Figure 1) was adopted from [24]. Firstly, 0.003 mol of iron(III) chloride was dissolved into 20 mL of ethylene glycol in a 150 mL glass beaker. Magnetic stirring at 50 °C for 10 min was applied. Sodium acetate (0.0122 mol) and PEG (0.5 g optional) were added into the solution under vigorous stirring (50 °C, 500 rpm), and the conditions were maintained until the materials were completely dissolved and the color of the solution became dirty yellow. A well-mixed solution was put into the microwave reactor (flexiWAVE, Milestone Srl, Milan, Italy), which was performing under 2.45 GHz of microwave irradiation. Temperature control was ensured by an optical fiber thermal sensor inserted into the glass tube with the reaction mixture. The tube was placed in a well-sealed Teflon vessel to maintain the pressure during the heating process. The reactor provided uniform sample heating to support the reaction, while the ethylene glycol with a higher dielectric constant was used as an energy adsorption agent. Continuous stirring and various times (from 1 to 120 min) and temperatures (200–250 °C) were used during the synthesis procedure. The temperature raising time was set to 5 (for a longer synthesis) or

2 (for a shorter synthesis) minutes. The longer time was chosen due to the higher stability of the temperature flux; however, when the synthesis time was shorter than 5 min, the temperature raising time was also reduced. After the synthesis, the solution became black, and the magnetic precipitates were collected using a neodymium magnet, washed with ethanol (three times) and distilled water (one time) and left to dry in an oven (120/300 LSN 11, SNOL, Utena, Lithuania).

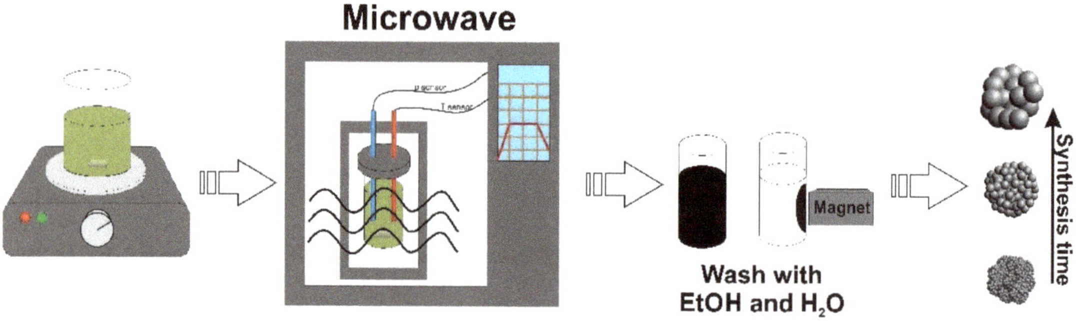

Figure 1. Synthesis scheme of Fe_3O_4 submicron spheres.

2.3. Characterization of Magnetic Fe_3O_4 Particles

The structures of the obtained magnetic particles were characterized using: XRD, Raman and infrared absorption spectroscopy, TEM and TGA. The magnetization of the particles was measured in a pulsed magnetic field by an induction method using well-compensated pick-up coils. For the characterization of the crystalline structure of the particles, XRD measurements were conducted using a MiniFlex (Rigaku, Tokyo, Japan) X-ray diffractometer with a scanning region of 2Θ from $10°$ to $80°$. The morphologies of the samples were studied using a high-resolution transmission electron microscope (HR-TEM Tecnai G2 F20 X-TWIN, FEI, Hillsboro, OR, USA). The infrared absorption (FTIR) spectra were recorded from particles dispersed in the KBr pellets using an Alpha spectrometer (Bruker, Inc., Karlsruhe, Germany) equipped with a room temperature RT-DLATGS detector. Some of the samples were measured by using the ATR accessory (Platinum ATR Diamond, Karlsruhe, Bruker). All of the FTIR spectra were collected with $4\ cm^{-1}$ resolution from 20 scans. The Raman spectra of the particles were obtained by an inVia Raman microscope (Renishaw, Wotton-under Edge, UK) equipped with a CCD camera thermoelectrically cooled to $-70\ °C$. The 830 nm laser excitation was restricted to 0.8 mW and was focused on the sample by a $20\times/0.4$ NA objective lens to a line-shape area on a sample of approximately $10\ \mu m \times 90\ \mu m$. The Raman signal was dispersed by using 830 lines/mm grating. Each sample was recorded at three different spots on its surface, with a 20 min acquisition time. The Raman wavenumber axis was calibrated by using the silicon standard peak at $520.7\ cm^{-1}$. The thermogravimetric analysis was performed using the STA6000 (Perkin Elmer, Waltham, MA, USA) with a 0.18 mL aluminum oxide crucible. The temperature was raised from $22\ °C$ to $500\ °C$ in a nitrogen atmosphere (20 mL/min). The temperature rising step was set to $10\ °C/min$. The magnetic characteristics at room temperature (294 K) were measured using pulsed-field magnetometry.

3. Results and Discussion

3.1. Investigation of Synthesis of Magnetic Particles

The syntheses of the magnetic particles (MPs) were performed at different time and temperature values. All of the used combinations are presented in Table 1. The temperature scale was set from $200°$ to $250\ °C$. A temperature of $180\ °C$ was also examined; however, even after 2 h of synthesis, black magnetic precipitates were still not visible. This might be due to the boiling point of ethylene glycol, which is $197\ °C$. Only the temperatures above the EG boiling point induce the reduction of Fe ions. It was noticed that various temperatures above the boiling point accelerate the particle formation reaction differently. For example, at 200 °C, the synthesis of MPs (without PEG in the synthesis mixture) results in black precipitates only after 90 min. If the temperature is raised to 250 °C, particles are formed even after 1 min in the microwave reactor (additionally, 2 min of rising temperature was set). The influence of the addition of polyethylene glycol (PEG) was also investigated. PEG is known to be a reductor for silver nanoparticles [41], so the reaction time of the formed Fe_3O_4 particles should also be affected. As could be seen in Table 1, this is true at low temperatures. At 200 °C, the reaction is completed after 30 min instead of 90 min without PEG. At 220 °C, with PEG, particles are obtained in 8 min, while without PEG, this would take 30 min. For a better understanding, a graph comparing the successful syntheses in the shortest time with and without PEG is presented (Figure 2). The influence of PEG is visible at low temperatures: 200–220 °C. However, at high temperatures (230–250 °C), no difference can be noticed, and the particles are obtained in a short (1–5 min) interval.

Table 1. Microwave-assisted solvothermal synthesis of Fe_3O_4 particles with and without PEG in the synthesis mixture [1].

Temp., °C Time, min	With PEG				Without PEG			
	200	220	230	250	200	220	230	250
1 [2]				+				+
2 [2]				+				+
5 [2]			+	+			+	+
8		+	+	+			+	+
10		+	+	+			+	
15		+	+	+			+	+
30	+	+	+	+		+		+
45	+							
60	+					+		
75	+	+						
90	+			+	+			+
105	+							
120				+	+	+		+

[1] Synthesis conducted at temperatures from 200 to 250 °C and from 1 to 120 min in the microwave reactor. The temperature rising time was set to 5 min unless stated otherwise. The successful reaction conditions (i.e., those that resulted in black magnetic precipitates) are marked with +. The table coloring indicates the conditions that yielded (green) and does not yield (blank) particles. [2] The temperature rising time is 2 min.

3.2. X-ray Diffraction Patterns

The crystal structure and phase purity of the obtained MPs were evaluated using XRD. The obtained patterns are shown in Figure 3A,B. All the obtained spectra were similar and matched the cubic Fe_3O_4 phase. The main obtained peaks—(220), (311), (400), (422), (511) and (440)—at 30.1°, 35.5°, 43.2°, 53.5°, 57.1° and 62.8° 2Θ values, respectively, match the COD (Crystallography Open Database) file, No. 9007644. At the most intense spectra, lower intensity peaks are also visible.

The intensity and the full width at half maximum (FWHM) of the diffraction peaks of the samples differ as well. In Figure 3A, the diffractograms of the particles synthesized at 250 °C are presented. For comparison, the syntheses with PEG and without PEG at synthesis times of 1 min and 30 min were chosen. All of them follow the XRD pattern for the magnetite structure; however, the intensity and FWHM of the spectra differ. Without PEG, the synthesized particles show higher intensity XRD spectra, while both spectra with PEG show lower intensity. It is known that the intensity or FWHM of the XRD peaks could be associated with the crystallite size. If the crystal structure is large, the obtained peaks are of a higher intensity, and small crystallites could show the low intensity of the XRD signal. From these data, particles with PEG are expected to be smaller in comparison to particles synthesized for the same time but without PEG.

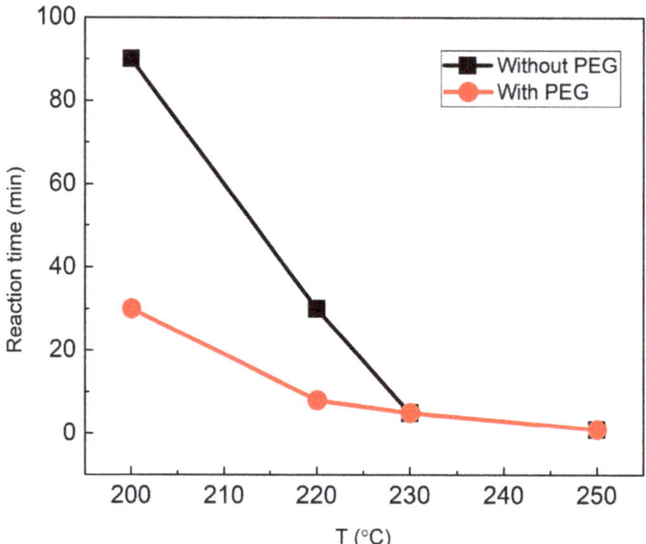

Figure 2. Reaction time dependence on the temperature maintained by the microwave reactor. A difference is noticed in the reaction times at low temperatures when PEG is added to the initial reaction mixture.

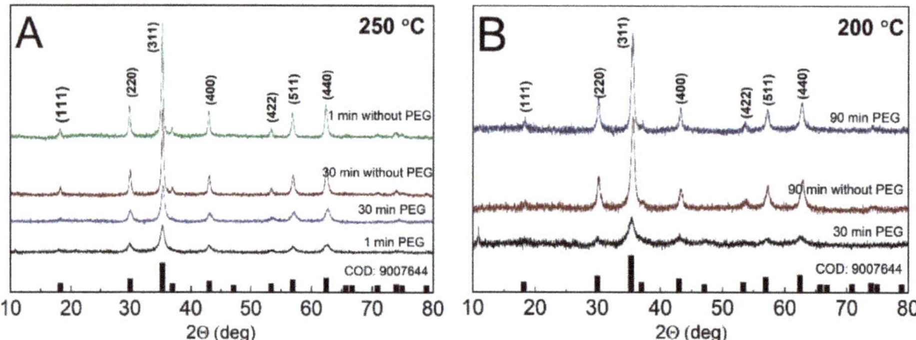

Figure 3. XRD diffractograms of Fe_3O_4 particles at synthesis temperatures of 250 °C (**A**) and 200 °C (**B**). The diffractograms of different synthesis times are presented, as well as a presence of PEG in the reaction mixture.

The XRD diffractograms were also compared for the samples synthesized at 200 °C. Figure 3B compares the particles obtained at the shortest possible synthesis times: 30 min with PEG and 90 min without PEG. The 90 min PEG-synthesis is also added for comparison. The diffractograms obtained at the same time—90 min—were of similar intensities; however, the intensity of the one obtained after 30 min of synthesis was the lowest. This, as well as the previous section, indicates that shorter synthesis times yield particles of a smaller crystallite structure, and, in time, the crystallites are growing, resulting in a more intensive XRD signal.

3.3. Raman Spectra Analysis

Magnetite and maghemite having the same spinel structure and almost identical lattice parameters make the identification of magnetite and maghemite by the XRD technique complicated [25,42]. However, Kozakova et al. suggested using the (511) Bragg peak (in the range of 56.5–57.5 of 2Θ) as an identification tool. For pure magnetite, the central position of the diffraction peak should be at 57.0°, while for maghemite, this peak is slightly shifted to higher values, i.e., 57.3° [25]. In our case the peak has a maximum at 57.1°, suggesting that the main structure of the sample is magnetite with some amount of maghemite. To analyze the crystal structure of magnetite and detect the possible secondary phases, Raman spectra were recorded. The literature suggests five Raman active vibrational modes at 193 cm^{-1} (T_{2g}), 306 cm^{-1} (E_g), 450–490 cm^{-1} (T_{2g}), 538 cm^{-1} (T_{2g}) and 668 cm^{-1} (A_{1g}) [43]. The characteristic peaks of magnetite in our measurements could be seen at 663–668 cm^{-1}, 308–310 cm^{-1} and close to 520 cm^{-1} for the samples prepared with and without PEG (Figure 4A,B). Raman spectra were recorded for the microwave heated samples from 1 to 90 min at 250 °C. The laser power density was reduced to 0.1 kW/cm^2 at the expense of longer acquisition times to ensure that no laser-induced photolytic and pyrolytic effects would take place in the sample [44,45]. Contrary to the synthesis without PEG, the 1 min microwave preparation at 250 °C with PEG resulted in MPs with distinctive narrow-bandwidth spectral modes at 245 and 376 cm^{-1} (Figure 4B). These low-wavenumber bands are associated with other secondary phases, most likely goethite and lepidocrocite [46], which were no longer present in the Raman spectra at increasingly longer heating times. The A_{1g} mode's asymmetry hinted at the presence of maghemite. Indeed, after fitting the experimental spectrum with Gaussian–Lorentzian shape components, the mode at 710 cm^{-1} was identified, which was directly associated with maghemite's A_{1g} mode. We estimated a 12–16% contribution to the total integral intensity from the maghemite at each tested microwave preparation with PEG. A stark difference can be seen in the A_{1g} mode's bandwidth expressed as the FWHM when the samples prepared with and without PEG are compared (Figure 4C). Generally, the FWHM correlates with the crystal structure of the sample and decreases with increasing crystallinity. It is well-known that the spectral modes of magnetite are much broader compared to, for example, those of hematite, due to the strong electron–phonon interactions [44,47]. However, the PEG preparation resulted in particles with bandwidths that were larger by 18 cm^{-1} on average compared to the ones prepared without PEG but with the same microwave heating time. The XRD data already confirmed larger crystallites in the nanoparticles prepared without PEG. A more quantitative analysis of the Raman bandwidths of the A_{1g} mode is provided in Figure 4C, where the FWHM is plotted against the sample preparation time at 250 °C. Our data show that, during the first 10 min of preparation without PEG, the FWHM increased from 56 to 66 cm^{-1}, which was followed by a sharp drop to 55 cm^{-1} at 15 min, with no significant change in subsequent heating. For the PEG preparation of the MPs, a change from 80 to 71 cm^{-1} was detected for the first 30 min; later on, the changes were marginal. For both preparations, the heating up to 30 min decreased the FWHM by ca. 10 cm^{-1}, indicating the growth of MPs with the increase in crystallinity in the samples with and without PEG.

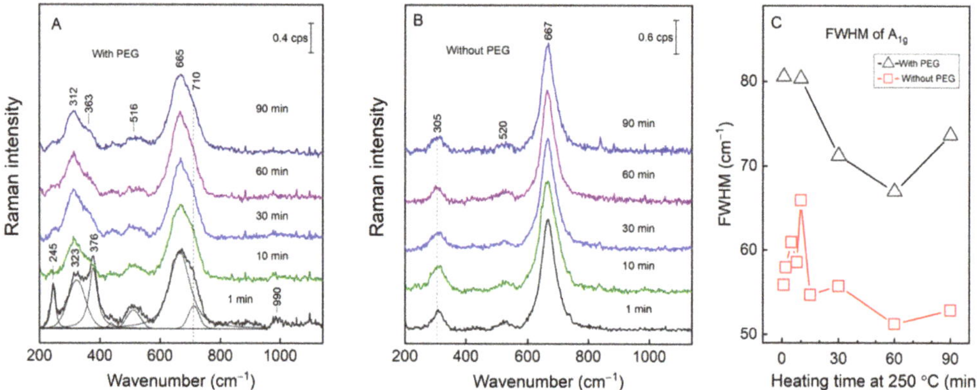

Figure 4. Raman spectra of the samples prepared with (**A**) and without (**B**) PEG heated from 90 to 1 min. The dependence of the full width at half maximum (FWHM) of the A_{1g} mode on the MPs preparation time at 250 °C with (black triangles) and without (red squares) PEG (**C**). The excitation wavelength was 830 nm; the laser power was set to 0.8 mW; the acquisition time was 60–120 min.

3.4. TEM Images Analysis

To confirm the different sizes of the crystallites, TEM images of the particles were captured. In Figure 5, the MPs obtained using synthesis without PEG at 220 °C and 250 °C are presented. These particles were synthesized at 220 °C for 120 (Figure 5A), 60 (Figure 5B) and 15 min (Figure 5C). Although the sizes of the particles are similar (around 200 nm (Table 2)), the structures of the MPs are quite different. The longest synthesized MPs are composed of larger sized crystallites (26 ± 4 nm). The synthesis that was 60 min in duration resulted in smaller crystallites (12 ± 2 nm), and the particles obtained after the shortest time of synthesis (15 min) had the smallest crystallites, the size of which was impossible to measure using the Image J program. The same results were also observed for the syntheses at higher temperatures. In Figure 5, the TEM images of the particles synthesized at 250 °C for 1 (Figure 5E) and 15 min (Figure 5D) are presented. The sizes of the spheres were more or less the same as those for the synthesis at 220 °C. Larger crystallites (25 ± 7 nm) were observed at longer synthesis times (15 min), while no observable crystallites were seen after 1 min of synthesis. The MP and crystallite sizes calculated by the Image J program are summarized in Table 2.

Table 2. Comparison of the Fe_3O_4 particle sizes synthesized without PEG at different temperatures and times. Particles refer to the whole sized spheres, and crystallites are the small grains of which the particles are comprised.

	220 °C			250 °C	
	15 min	**60 min**	**120 min**	**1 min**	**15 min**
Particles, nm	188 ± 27	182 ± 45	229 ± 55	303 ± 134	200 ± 29
Crystallites, nm	NA [1]	12 ± 2	26 ± 4	NA [1]	25 ± 7

[1] Data is not available to obtain using ImageJ programe.

3.5. FTIR Spectra Analysis

To evaluate the presence of organic reductors on the MPs, FTIR spectra were recorded for the particles prepared with and without PEG (Figure 6). Magnetite, due to its spinal structure, has four infrared-active bands which appear at ca. 570 (ν_1), 390 (ν_2), 270 (ν_3) and 180 (ν_4) cm^{-1} [48–51]. The strong ν_1 mode assigned to the Fe–O stretching motion of the tetrahedral and octahedral sites, when narrow, suggests the high crystallinity of the sample. In the case of the formation of defects and secondary phases, the modes become broader and shift. For maghemite, which is considered to be a defective form of magnetite, the Fe–O stretch absorption modes are expected at 630, 590 and 430 cm^{-1} [48,52]. We already discussed the relatively small contribution from γ-Fe$_2$O$_3$ to the particles prepared with PEG and the nonexistent contribution for the MPs prepared without PEG based on our Raman measurements. Therefore, the relatively broad infrared absorption feature near 600 cm^{-1} was ascribed to the Fe$_3$O$_4$ phase in nanoparticles of low crystallinity and, to some extent, to γ-Fe$_2$O$_3$. Notably, the preparation without PEG resulted in MPs with a somewhat narrower mode—near 600 cm^{-1}. This is arguably due to the higher crystallinity of the MPs compared with the PEG MPs. The most obvious heating-time-dependent changes occurred within the first 10 min. These are especially visible in the 800–1100 cm^{-1} region, where the characteristic vibrations of the organic reductors could be seen. For example, the modes near 880, 1040 and 1080 cm^{-1} correspond to the vibrations of ethylene glycol and PEG (Figure 6C). Later, as the heating time passed the 10 min mark, the spectral changes in the PEG preparation were marginal, indicating that the reducing reagents had fully reacted and that no further changes had happened with the organic components in the reaction vessel.

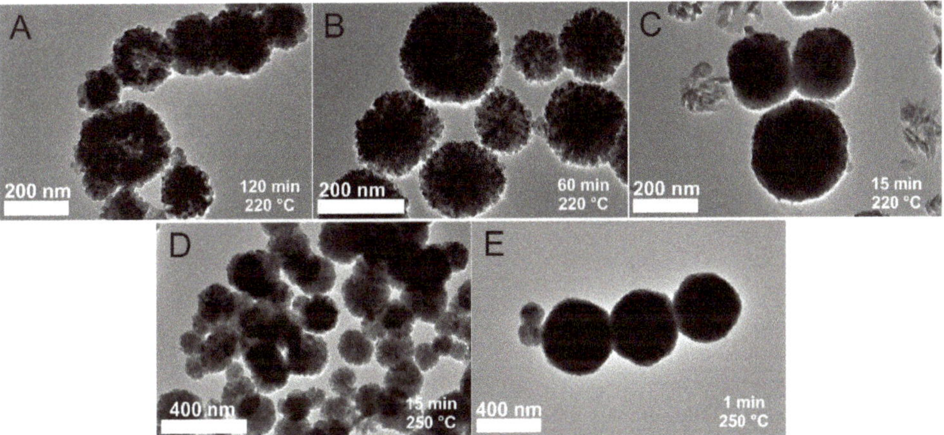

Figure 5. TEM images of the synthesized samples without PEG: 220 °C, 120 min (**A**); 220 °C, 60 min (**B**); 220 °C, 15 min (**C**); 250 °C, 15 min (**D**); 250 °C, 1 min (**E**). The scale bar for (**A–C**) is 100 nm, and the scale bar for (**D**) and (**E**) is 200 nm.

3.6. Thermo Gavimetric Analysis

In Figure 7, the thermogravimetric analysis data of weight change and heat flow are presented. Few temperature intervals could be detected in the following samples: from RT to 200 °C, from 200 to 350 °C and from 350 to 500 °C. In the first interval, the endothermic loss of water and -OH groups are detected. The curve decrease is small, reaching up to 2% of the weight loss for the samples synthesized with PEG, while for the ones without PEG, the loss is 1% or even nonexistent. The second interval could be attributed to the desorption and subsequent evaporation of PEG or EG and the last interval—the phase transformation from Fe$_3$O$_4$ to γ-Fe$_2$O$_3$. The evaporation of PEG and EG resulted in a larger decrease in weight in comparison to that of the first interval. The samples with PEG lost 4–6% of their weight during the second interval, while the samples without PEG lost 1–4% of their weight. The highest amount of remnant organic material within the MPs prepared for 1 min was already demonstrated, as seen in the FTIR data. In the last interval, the weight of the samples remains similar—except for the sample synthesized without PEG for 30 min. Here, the increase of the weight at about 1% is registered. The phase transformation from Fe$_3$O$_4$ to γ-Fe$_2$O$_3$ is reached and then additional oxygen is introduced to the magnetite crystal structure. The process for the sample obtained after 30 min of synthesis without

PEG is probably the most efficient, so the mass increase is registered. Although the mass of the sample stayed similar at the third interval, the heat flow was increasing, suggesting the occurrence of the exothermic process. The start of this process for all the samples begins at around 250–300 °C at the second interval. Possibly, the exothermic oxidation from Fe(II) to Fe(III) starts earlier and overlaps with the endothermic evaporation of the organics.

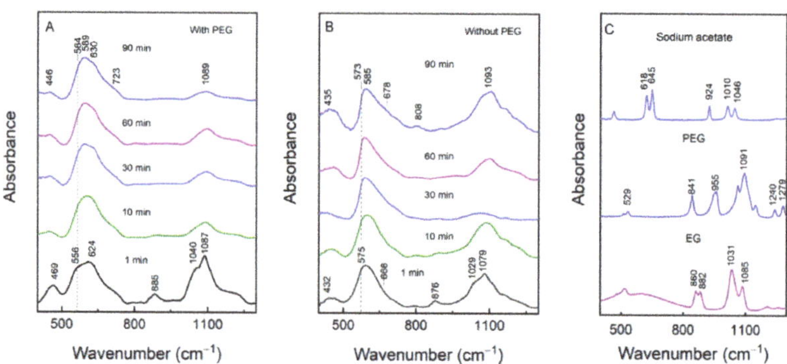

Figure 6. Preparation time at 250 °C-dependent FTIR spectra of KBr dispersed particles that were produced with (**A**) and without (**B**) PEG. The ATR-FTIR spectra of sodium acetate, PEG and ethylene glycol (EG) (**C**). The spectral positions in (**A**) and (**B**) were worked out from the second derivative spectrum calculated using the Savitzky–Golay algorithm.

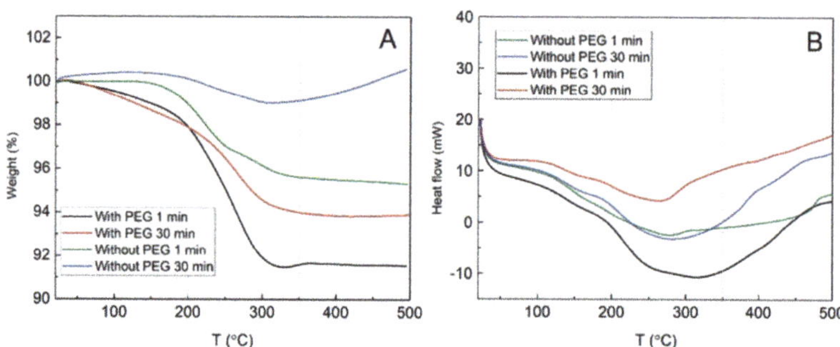

Figure 7. TGA data of samples synthesized with PEG and without PEG; synthesis time—1 or 30 min at 250 °C. Changes in sample weight (**A**) and heat flow (**B**) are registered from room temperature to 500 °C.

3.7. Investigation of Magnetic Properties

The magnetization M of particles was measured in a pulsed magnetic field with a duration of about 4 ms using a pulse magnetizer. The advantage of this method is that it allows for the rapid acquisition of isothermal magnetization, and a higher magnetic field can be applied to the samples. This method is widely used for measuring the magnetization of strong magnets [53], ferromagnetic materials or superconductors [54] and ferromagnetic powder or volcanic rock [55]. Moreover, it was demonstrated by Kodama [56] that the measurement of the magnetization of nanoparticles using this method is acceptable and yields results similar to those yielded by the vibration method. It was shown that a small difference in the obtained results is caused and phenomenologically explained by the difference in the time scale of the magnetization processes under consideration. That is, in the case of the pulsed field, a fraction of the magnetic particles with a relaxation time longer than the pulse rise time fails to follow the pulse. For the vibration method, the field sweep rate is about four orders of magnitude slower than the pulsed field duration, so most of the magnetic particles are magnetized simultaneously with the applied field.

In our work, all of the magnetic characteristics of the MPs using this method were compared at room temperature (294 K). The measurement system consists of a pulse magnetizer and two coils positioned inside of this magnetizer. The coils were connected with each other in opposite directions and were well-compensated. The coil of the pulse magnetizer with an inner diameter of 2.5 cm connected to the capacitor of 80 μF generates a pulsed magnetic field with an amplitude of 1.2 kOe and a pulse duration of 4 ms. The signal, directly proportional to M vs. time derivative, was obtained when the sample was placed in one of the coils and the capacitor was discharged through the pulse magnetizer. For the magnetic field measurements, the additional pick-up coil system was used. The saturations of mass magnetization (M_S), coercivity (H_C) and remanent magnetization (M_r) were measured in this case.

The nonlinear magnetization curves with the hysteresis loop, characteristic of the ferromagnetic behavior, are clearly observed in all the samples (Figure 8A,B). It was found, that, for the samples synthesized without PEG (see Figure 8A), the saturation magnetization M_S increases with the increase in the synthesis time of the MPs. The magnetization of the particles that have been synthesized with PEG (Figure 8B) shows the same tendency, but the saturation value is lower than that for those that have been synthesized without PEG. The coercive field for all of the samples also depends on the synthesis method. The summarized results of M_S and H_C versus synthesis time are shown in Figure 8C. It can be seen that, for all of the samples, the increase in the synthesis time leads to an increase in the saturation magnetization and coercive field. For example, for the samples synthesized without PEG, the saturation magnetization increased from 32 emu/g to 78 emu/g when the synthesis time of the particles was changed from 1 min to 30 min. Meanwhile, the samples synthesized with PEG show a reduced value of M_S for a longer times of synthesis, and it is changed from 46 emu/g to 62 emu/g, respectively. The lower magnetization can be attributed to the presence of a magnetically disordered layer or the existence of a secondary phase in these particles.

It is known that the magnetic properties of iron-oxide nanoparticles strongly depend on the particle size, shape and composition. Moreover, the synthesis method influences the stoichiometry of the nanoparticles, i.e., iron oxides can be synthesized in two main phases: magnetite (Fe_3O_4) or maghemite (γ-Fe_2O_3) [25,35,42]. Furthermore, it is known that MPs formed from magnetite have a much higher saturation magnetization than those formed from maghemite [57,58]. An analysis of the Raman spectra and data of the Fourier-transform infrared spectroscopy shows that the samples synthesized with PEG have a mixture of magnetite and maghemite, while for the samples without PEG, only magnetite is present. These results are also confirmed by the measurements of the particles' magnetization. Moreover, the $M(H)$ curve gives information about the domain structure in an ensemble of the MPs. It is well-known that magnetite nanoparticles that are smaller than 100 nm are in a single-domain state with a relatively low magnetization. An increase in particle size leads to a multidomain structure and an increased saturation magnetization [58–60]. Our results are in good agreement with the literature data. Table 2 shows that an increase in the synthesis time leads to an increase in the size of the crystallites, which are small grains in the submicron sized spheres.

The other magnetization parameter that was measured is the coercivity of the samples. It can be seen from Figure 8, that, for the samples prepared with PEG, the coercivity increases monotonously from 2 mT to 2.8 mT when the time of the synthesis is changed from 1 to 30 min, while for the samples prepared without PEG, it first decreases from 4.5 to 3 mT (until 10 min) and later increases up to 5.5 mT. These results could be explained by the peculiarity of the crystalline structures of MPs and also by the increase in their sizes. It was shown by Dehsari et al. [58] that the behavior of H_C with the size is commonly attributed to the transition of the particle from a magnetically single-domain to a multidomain structure.

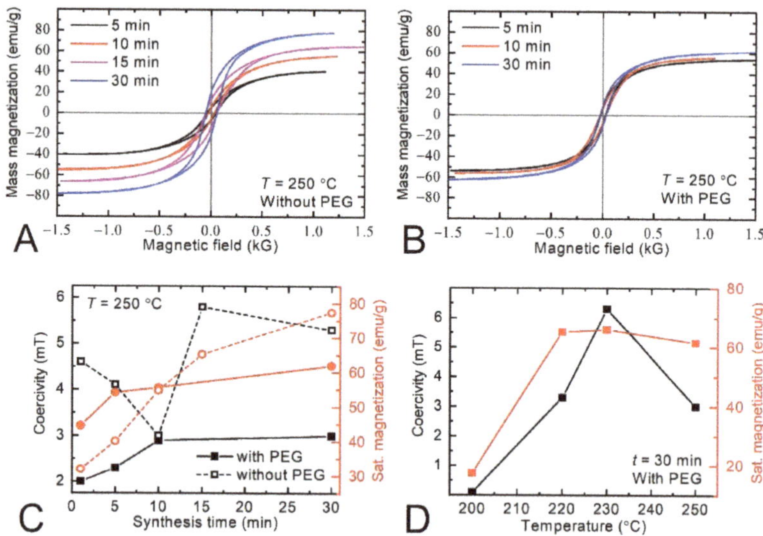

Figure 8. Hysteresis loop of the samples synthesized without PEG at 250 °C for different synthesis times—5, 10, 15 and 30 min—(**A**) and that of the samples synthesized with PEG (**B**). Magnetization (black symbols) and coercivity (red symbols) of the samples synthesized with PEG (solid symbols) and without PEG (empty symbols) using different times, from 1 to 30 min at 250 °C (**C**). Saturation magnetization (red symbols) and coercivity (black symbols) of the samples synthesized with PEG at different temperatures (**D**).

In addition, the obtained results of the magnetization of the samples were analyzed from the point of view of the dependence of their parameters on the synthesis temperature. The obtained results are shown in the Figure 8D. As can be seen from the graph, the magnetization and coercivity of the samples increase with the increase in synthesis temperature from 200 °C to 230 °C, but at higher temperatures, these parameters slightly decrease. However, this is related mostly to the peculiarities of the synthesis of the particles, not to their magnetic properties.

3.8. The Mechanism of Magnetite Formation

According to the literature, the microwave-assisted solvothermal synthesis of magnetite particles could be divided into two stages. The first is called the nucleation of primary crystals and the second is called the nanoparticles aggregation [27,61]. In the synthesis mixture, sodium acetate acts as a weak base, helping the EG to reduce iron ions, and, in the presence of the trace amount of water, it can be hydrolyzed and release OH^- ions. The trace amount of water could be obtained from the air as the initial mixture is stirred in an ambient atmosphere, and, in EG, about 0.5% (w/w) is water. Only the trace amount of water is required for the synthesis. The addition of extra water results in polyhedral particles of different sizes [25].

The hydrolysis of sodium acetate proceeds as follows (the net ionic equation is presented):

$$CH_3COO^- + H_2O \rightleftarrows CH_3COOH + OH^- \tag{1}$$

The OH^- ions are consumed for Fe(III)hydroxide formation:

$$Fe^{3+} + 3OH^- \rightleftarrows Fe(OH)_3 \tag{2}$$

which later may turn to Fe_2O_3:

$$2Fe(OH)_3 \rightleftharpoons Fe_2O_3 + 3H_2O \qquad (3)$$

In the meantime, ethylene glycol can undergo dehydration and form acetaldehyde [62]:

$$2HOCH_2 - CH_2OH \rightleftharpoons 2CH_3CHO + 2H_2O \qquad (4)$$

Then, the acetaldehyde reacts with Fe^{3+} ions and reduces them to Fe^{2+}:

$$2CH_3CHO + 2Fe^{3+} \rightleftharpoons CH_3COCOCH_3 + 2Fe^{2+} + 2H^+ \qquad (5)$$

Additionally, an alternative pathway of EG may also exist. For example, the heating of EG in air may generate glycolaldehyde, a reductant for many metal ions [63]:

$$2HOCH_2CH_2OH + O_2 \rightleftharpoons 2HOCH_2CHO + 2H_2O \qquad (6)$$

In any case, the obtained Fe^{2+} forms hydroxide:

$$Fe^{2+} + 2OH^- \rightleftharpoons Fe(OH)_2 \qquad (7)$$

Finally, in the presence of both Fe ions, magnetite formation is enabled:

$$Fe(OH)_2 + 2Fe(OH)_3 \rightleftharpoons Fe_3O_4 + 4H_2O \qquad (8)$$

The addition of microwaves is believed to facilitate hydroxide formation to oxide reaction (8), EG dehydration and subsequent reactions (due to the energy absorption and heating up of the solvent). In addition to this, secondary aggregation to the submicron size spheres is also believed to be caused by microwaves [27].

In the presence of PEG, more Fe_2O_3 crystalline structures are left. Possibly, the long molecules of PEG limit the diffusion of ions, and the first part of the reaction mechanism (1–3) is dominating. However, PEG is known to act as an additional reductor as well [41].

4. Conclusions

In this work, a facile and eco-friendly microwave-assisted solvothermal method is suggested for the synthesis of Fe_3O_4 magnetite spheres. Depending on the reaction temperature, the minimal time is suggested for fully synthesized MPs. At the temperature close to the ethylene glycol boiling point (200 °C), the shortest synthesis time is 90 min for the preparation without PEG and 30 min if PEG is used in the initial synthesis solution. However, when the temperature is increased to 250 °C, the fully synthesized magnetic particles are obtained even after 1 min of reaction (with an additional 2 min temperature raising time), independently of the presence of PEG. Although the sizes of the spheres at different synthesis times are similar, the crystal structures of these spheres differ. The longer the synthesis time is, the larger the obtained crystals are. These results were confirmed by TEM and XRD measurements. From the FTIR and Raman measurements, the sample synthesized with PEG contains a mixture of magnetite and maghemite, while for the samples without PEG, only magnetite is present. The magneticity measurement results also confirm this statement. It was found that the saturation magnetization and coercive field increase with the increase in synthesis time. We hope that this research will be beneficial for the further synthesis, development and applications of magnetite particles.

Author Contributions: Synthesis, methodology and XRD measurements, G.Z.; Raman and FTIR measurements and interpretation, writing—review and editing, M.T.; TEM visualization, review and editing A.D.; Magneticity measurements, interpretation, review and editing J.D. and V.S.; Conceptualization, writing—original draft preparation, systematization of the results, project administration L.M.; Supervision, writing—review and editing, G.N. All authors have read and agreed to the published version of the manuscript.

Funding: This research is funded by the European Social Fund under the No 09.3.3-LMT-K-712-19-0142 "Development of Competences of Scientists, other Researchers and Students through Practical Research Activities" measure under a grant agreement with the Research Council of Lithuania (LMTLT).

Institutional Review Board Statement: Not applicable.

Informed Consent Statement: Not applicable.

Data Availability Statement: The data presented in this study are available from the corresponding authors upon reasonable request.

Acknowledgments: The authors would like to acknowledge Milita Vagner (Vilnius University) for the TGA measurements and Simas Sakirzanovas (Vilnius University) for the technical support.

Conflicts of Interest: The authors declare that they have no known competing financial interest or personal relationships that could have appeared to influence the work reported in this paper.

References

1. Materón, E.M.; Miyazaki, C.M.; Carr, O.; Joshi, N.; Picciani, P.H.S.; Dalmaschio, C.J.; Davis, F.; Shimizu, F.M. Magnetic Nanoparticles in Biomedical Applications: A Review. *Appl. Surf. Sci. Adv.* **2021**, *6*, 100163. [CrossRef]
2. Ali, A.; Shah, T.; Ullah, R.; Zhou, P.; Guo, M.; Ovais, M.; Tan, Z.; Rui, Y.K. Review on Recent Progress in Magnetic Nanoparticles: Synthesis, Characterization, and Diverse Applications. *Front. Chem.* **2021**, *9*, 548. [CrossRef] [PubMed]
3. Farzin, A.; Etesami, S.A.; Quint, J.; Memic, A.; Tamayol, A. Magnetic Nanoparticles in Cancer Therapy and Diagnosis. *Adv. Healthcare Mater.* **2020**, *9*, 1901058. [CrossRef] [PubMed]
4. Reddy, L.H.; Arias, J.L.; Nicolas, J.; Couvreur, P. Magnetic Nanoparticles: Design and Characterization, Toxicity and Biocompatibility, Pharmaceutical and Biomedical Applications. *Chem. Rev.* **2012**, *112*, 5818–5878. [CrossRef] [PubMed]
5. Dilnawaz, F.; Singh, A.; Mohanty, C.; Sahoo, S.K. Dual Drug Loaded Superparamagnetic Iron Oxide Nanoparticles for Targeted Cancer Therapy. *Biomaterials* **2010**, *31*, 3694–3706. [CrossRef] [PubMed]
6. Zhao, Y.; Qiu, Z.; Huang, J. Preparation and Analysis of Fe3O4 Magnetic Nanoparticles Used as Targeted-Drug Carriers. *Chin. J. Chem. Eng.* **2008**, *16*, 451–455. [CrossRef]
7. Leong, S.S.; Ahmad, Z.; Low, S.C.; Camacho, J.; Faraudo, J.; Lim, J.K. Unified View of Magnetic Nanoparticle Separation under Magnetophoresis. *Langmuir* **2020**, *36*, 8033–8055. [CrossRef]
8. Liu, X.; Zhang, Y.; Wang, Y.; Zhu, W.; Li, G.; Ma, X.; Zhang, Y.; Chen, S.; Tiwari, S.; Shi, K.; et al. Comprehensive Understanding of Magnetic Hyperthermia for Improving Antitumor Therapeutic Efficacy. *Theranostics* **2020**, *10*, 3793–3815. [CrossRef]
9. Di, S.; Ning, T.; Yu, J.; Chen, P.; Yu, H.; Wang, J.; Yang, H.; Zhu, S. Recent Advances and Applications of Magnetic Nanomaterials in Environmental Sample Analysis. *Trends Analyt Chem.* **2020**, *126*, 115864. [CrossRef]
10. Sappino, C.; Primitivo, L.; de Angelis, M.; Domenici, M.O.; Mastrodonato, A.; Romdan, I.B.; Tatangelo, C.; Suber, L.; Pilloni, L.; Ricelli, A.; et al. Functionalized Magnetic Nanoparticles as Catalysts for Enantioselective Henry Reaction. *ACS Omega* **2019**, *4*, 21809–21817. [CrossRef]
11. Haun, J.B.; Yoon, T.J.; Lee, H.; Weissleder, R. Magnetic Nanoparticle Biosensors. *Wiley Interdiscip. Rev. Nanomed. Nanobiotechnol.* **2010**, *2*, 291–304. [CrossRef] [PubMed]
12. Khorsand Zak, A.; Shirmahd, H.; Mohammadi, S.; Banihashemian, S.M. Solvothermal Synthesis of Porous Fe₃O₄ Nanoparticles for Humidity Sensor Application. *Mater. Res. Express* **2020**, *7*, 025001. [CrossRef]
13. Lai, H.; Xu, F.; Wang, L.A. Review of the Preparation and Application of Magnetic Nanoparticles for Surface-Enhanced Raman Scattering. *J. Mater. Sci.* **2018**, *53*, 8677–8698. [CrossRef]
14. Michałowska, A.; Krajczewski, J.; Kudelski, A. Magnetic Iron Oxide Cores with Attached Gold Nanostructures Coated with a Layer of Silica: An Easily, Homogeneously Deposited New Nanomaterial for Surface-Enhanced Raman Scattering Measurements. *Spectrochim. Acta A Mol. Biomol. Spectrosc.* **2022**, *277*, 121266. [CrossRef]
15. Michałowska, A.; Żygieło, M.; Kudelski, A. Fe3O4-Protected Gold Nanoparticles: New Plasmonic-Magnetic Nanomaterial for Raman Analysis of Surfaces. *Appl. Surf. Sci.* **2021**, *562*, 150220. [CrossRef]
16. Ge, S.; Shi, X.; Sun, K.; Li, C.; Uher, C.; Baker, J.R.; Banaszak Holl, M.M.; Orr, B.G. Facile Hydrothermal Synthesis of Iron Oxide Nanoparticles with Tunable Magnetic Properties. *J. Phys. Chem. C* **2009**, *113*, 13593–13599. [CrossRef]
17. Chen, Y.; Zhang, J.; Wang, Z.; Zhou, Z. Solvothermal Synthesis of Size-Controlled Monodispersed Superparamagnetic Iron Oxide Nanoparticles. *Appl. Sci.* **2019**, *9*, 5157. [CrossRef]
18. Unni, M.; Uhl, A.M.; Savliwala, S.; Savitzky, B.H.; Dhavalikar, R.; Garraud, N.; Arnold, D.P.; Kourkoutis, L.F.; Andrew, J.S.; Rinaldi, C. Thermal Decomposition Synthesis of Iron Oxide Nanoparticles with Diminished Magnetic Dead Layer by Controlled Addition of Oxygen. *ACS Nano* **2017**, *11*, 2284–2303. [CrossRef]
19. Tsuji, M.; Hashimoto, M.; Nishizawa, Y.; Kubokawa, M.; Tsuji, T. Microwave-Assisted Synthesis of Metallic Nanostructures in Solution. *Chem. Eur. J.* **2005**, *11*, 440–452. [CrossRef]

20. Zhu, X.H.; Hang, Q.M. Microscopical and Physical Characterization of Microwave and Microwave-Hydrothermal Synthesis Products. *Micron* **2013**, *44*, 21–44. [CrossRef]

21. Xu, J.; Yang, H.; Fu, W.; Du, K.; Sui, Y.; Chen, J.; Zeng, Y.; Li, M.; Zou, G. Preparation and Magnetic Properties of Magnetite Nanoparticles by Sol-Gel Method. *J. Magn. Magn. Mater.* **2007**, *309*, 307–311. [CrossRef]

22. Lemine, O.M.; Omri, K.; Zhang, B.; el Mir, L.; Sajieddine, M.; Alyamani, A.; Bououdina, M. Sol-Gel Synthesis of 8 nm Magnetite (Fe3O4) Nanoparticles and Their Magnetic Properties. *Superlattices Microstruct.* **2012**, *52*, 793–799. [CrossRef]

23. Tai, M.F.; Lai, C.W.; Hamid, S.B.A.; Suppiah, D.D.; Lau, K.S.; Yehya, W.A.; Julkapli, N.M.; Lee, W.H.; Lim, Y.S. Facile Synthesis of Magnetite Iron Oxide Nanoparticles via Precipitation Method at Different Reaction Temperatures. *Mater. Res. Innov.* **2014**, *18*, S6-470–S6-473. [CrossRef]

24. Juang, R.-S.; Su, C.-J.; Wu, M.-C.; Lu, H.-C.; Wang, S.-F.; Sun, A.-C. Fabrication of Magnetic Fe$_3$O$_4$ Nanoparticles with Unidirectional Extension Pattern by a Facile and Eco-Friendly Microwave-Assisted Solvothermal Method. *J. Nanosci. Nanotechnol.* **2019**, *19*, 7645–7653. [CrossRef]

25. Kozakova, Z.; Kuritka, I.; Kazantseva, N.E.; Babayan, V.; Pastorek, M.; Machovsky, M.; Bazant, P.; Saha, P. The Formation Mechanism of Iron Oxide Nanoparticles within the Microwave-Assisted Solvothermal Synthesis and Its Correlation with the Structural and Magnetic Properties. *Dalton Trans.* **2015**, *44*, 21099–21108. [CrossRef]

26. Hernández-Hernández, A.A.; Álvarez-Romero, G.A.; Castañeda-Ovando, A.; Mendoza-Tolentino, Y.; Contreras-López, E.; Galán-Vidal, C.A.; Páez-Hernández, M.E. Optimization of microwave-solvothermal synthesis of Fe$_3$O$_4$ nanoparticles. Coating, modification, and characterization. *Mater. Chem. Phys.* **2018**, *205*, 113–119. [CrossRef]

27. Zanchettin, G.; Falk, G.S.; González, S.Y.G.; Hotza, D. High performance magnetically recoverable Fe$_3$O$_4$ nanocatalysts: Fast microwave synthesis and photo-fenton catalysis under visible-light. *Chem. Eng. Process.* **2021**, *166*, 108438. [CrossRef]

28. Bhattacharjee, S.; Mazumder, N.; Mondal, S.; Panigrahi, K.; Banerjee, A.; Das, D.; Sarkar, S.; Roy, D.; Chattopadhyay, K.K. Size-modulation of functionalized Fe$_3$O$_4$: Nanoscopic customization to devise resolute piezoelectric nanocomposites. *Dalton Trans.* **2020**, *49*, 7872–7890. [CrossRef]

29. Jing, X.; Liu, T.; Wang, D.; Liu, J.; Meng, L. Controlled synthesis of water-dispersive and superparamagnetic Fe$_3$O$_4$ nanomaterials by a microwave-asisted solvothermal method: From nanocrystals to nanoclusters. *CrystEngComm* **2017**, *19*, 5089–5099. [CrossRef]

30. Rizzuti, A.; Dissisti, M.; Mastrorilli, P.; Sportelli, M.C.; Cioffi, N.; Picca, R.A.; Agostinelli, E.; Varvaro, G.; Caliandro, R. Shape-control by microwave-assisted hydrothermal method for the synthesis of magnetite nanoparticles using organic additives. *J. Nanopart. Res.* **2015**, *17*, 408. [CrossRef]

31. Liang, Y.J.; Zhang, Y.; Guo, Z.; Xie, J.; Bai, T.; Zou, J.; Gu, N. Ultrafast Preparation of Monodisperse Fe$_3$O$_4$ Nanoparticles by Microwave-Assisted Thermal Decomposition. *Chem. Eur. J.* **2016**, *22*, 11807–11815. [CrossRef] [PubMed]

32. Matijevic, E. Preparation and Properties of Uniform Size Colloids. *Chem. Mater.* **1993**, *5*, 412–426. [CrossRef]

33. Li, X.; Zhang, F.; Ma, C.; Saul, E.; He, N. Green Synthesis of Uniform Magnetite (Fe3O4) Nanoparticles and Micron Cubes. *J. Nanosci. Nanotechnol.* **2012**, *12*, 2939–2942. [CrossRef] [PubMed]

34. Li, Q.; Kartikowati, C.W.; Horie, S.; Ogi, T.; Iwaki, T.; Okuyama, K. Correlation between Particle Size/Domain Structure and Magnetic Properties of Highly Crystalline Fe3O4 Nanoparticles. *Sci. Rep.* **2017**, *7*, 9894. [CrossRef]

35. Teja, A.S.; Koh, P.Y. Synthesis, Properties, and Applications of Magnetic Iron Oxide Nanoparticles. *Prog. Cryst. Growth Charact. Mater.* **2009**, *55*, 22–45. [CrossRef]

36. Mascolo, M.C.; Pei, Y.; Ring, T.A. Room Temperature Co-Precipitation Synthesis of Magnetite Nanoparticles in a Large pH Window with Different Bases. *Materials* **2013**, *6*, 5549–5567. [CrossRef]

37. Rahmayanti, M. Synthesisof Magnetite Nanoparticles Using Reverse Co-Precipitation Method With NH$_4$OH as Precipitating Agent and Its Stability Test at Various pH. *Nat. Sci.* **2020**, *9*, 54–58. [CrossRef]

38. Laurent, S.; Forge, D.; Port, M.; Roch, A.; Robic, C.; vander Elst, L.; Muller, R.N. Magnetic Iron Oxide Nanoparticles: Synthesis, Stabilization, Vectorization, Physicochemical Characterizations and Biological Applications. *Chem. Rev.* **2008**, *108*, 2064–2110. [CrossRef]

39. Castelló, J.; Gallardo, M.; Busquets, M.A.; Estelrich, J. Chitosan (or Alginate)-Coated Iron Oxide Nanoparticles: A Comparative Study. *Colloids Surf. A: Physicochem. Eng. Asp.* **2015**, *468*, 151–158. [CrossRef]

40. Harraz, F.A. Polyethylene Glycol-Assisted Hydrothermal Growth of Magnetite Nanowires: Synthesis and Magnetic Properties. *Phys. E Low-Dimens. Syst. Nanostructures* **2008**, *40*, 3131–3136. [CrossRef]

41. Luo, C.; Zhang, Y.; Zeng, X.; Zeng, Y.; Wang, Y. The Role of Poly(Ethylene Glycol) in the Formation of Silver Nanoparticles. *J. Colloid Interface Sci.* **2005**, *288*, 444–448. [CrossRef] [PubMed]

42. Alibeigi, S.; Vaezi, M.R. Phase Transformation of Iron Oxide Nanoparticles by Varying the Molar Ratio of Fe^{2+}:Fe^{3+}. *Chem. Eng. Technol.* **2008**, *31*, 1591–1596. [CrossRef]

43. Testa-Anta, M.; Ramos-Docampo, M.A.; Comesaña-Hermo, M.; Rivas-Murias, B.; Salgueiriño, V. Raman Spectroscopy to Unravel the Magnetic Properties of Iron Oxide Nanocrystals for Bio-Related Applications. *Nanoscale Adv.* **2019**, *1*, 2086–2103. [CrossRef]

44. Shebanova, O.N.; Lazor, P. Raman Study of Magnetite (Fe3O4): Laser-Induced Thermal Effects and Oxidation. *J. Raman Spectrosc.* **2003**, *34*, 845–852. [CrossRef]

45. Jubb, A.M.; Allen, H.C. Vibrational Spectroscopic Characterization of Hematite, Maghemite, and Magnetite Thin Films Produced by Vapor Deposition. *ACS Appl. Mater. Interfaces* **2010**, *2*, 2804–2812. [CrossRef]

46. Li, S.; Hihara, L.H. A Micro-Raman Spectroscopic Study of Marine Atmospheric Corrosion of Carbon Steel: The Effect of Akaganeite. *J. Electrochem. Soc.* **2015**, *162*, C495–C502. [CrossRef]
47. Gupta, R.; Sood, A.K.; Metcalf, P.; Honig, J.M. Raman Study of Stoichiometric and Zn-Doped Fe3O4. *Phys. Rev. B Condens. Matter* **2002**, *65*, 1044301–1044308. [CrossRef]
48. Nasrazadani, S.; Raman, A. The Application of Infrared Spectroscopy to the Study of Rust Systems-II. Study of Cation Deficiency in Magnetite (Fe_3O_4) Produced During its Transformation to Maghemite (γ-Fe_2O_3) And Hematite (α-Fe_2O_3). *Corros. Sci.* **1993**, *34*, 1355–1365. [CrossRef]
49. Ishii, M.; Nakahira, M. Infrared Absorption Spectra and Cation Distributions in (Mn, Fe)$_3$O$_4$. *Solid State Commun.* **1972**, *11*, 209–212. [CrossRef]
50. Stoia, M.; Istratie, R.; Păcurariu, C. Investigation of Magnetite Nanoparticles Stability in Air by Thermal Analysis and FTIR Spectroscopy. *J. Therm. Anal. Calorim.* **2016**, *125*, 1185–1198. [CrossRef]
51. Wang, Q.; Shi, J.L.; Chen, L.D.; Yan, D.S. Synthesis of Nanocrystalline Magnetite (Fe_3O_4) Films by Self-Reduction Sol-Gel Route. *Mater. Sci. Forum* **2003**, *423–424*, 569–572. [CrossRef]
52. Namduri, H.; Nasrazadani, S. Quantitative Analysis of Iron Oxides Using Fourier Transform Infrared Spectrophotometry. *Corros. Sci.* **2008**, *50*, 2493–2497. [CrossRef]
53. Nakahata, Y.; Borkowski, B.; Shimoji, H.; Yamada, K.; Todaka, T.; Enokizono, M. Precise Measurement of Magnetization Characteristics in High Pulsed Field. *J. Appl. Phys.* **2012**, *111*, 07A712. [CrossRef]
54. Trojanowski, S.; Ciszek, M. Sensitivity of the Integrating Pulse Magnetometer with a First-order Gradiometer. *Rev. Sci. Instr.* **2008**, *79*, 104702. [CrossRef] [PubMed]
55. Kodama, K. Pulsed-field Magnetometry for Rock Magnetism. *Earth Planet Sp.* **2015**, *67*, 122. [CrossRef]
56. Kodama, K. Measurement of Dynamic Magnetization Induced by a Pulsed Field: Proposal for a New Rock Magnetism Method. *Front. Earth Sci.* **2015**, *3*, 5. [CrossRef]
57. Aivazoglou, E.; Metaxa, E.; Hristoforou, E. Microwave-Assisted Synthesis of Iron Oxide Nanoparticles in Biocompatible Organic Environment. *AIP Adv.* **2018**, *8*, 048201. [CrossRef]
58. Dehsari, H.S.; Ksenofontov, V.; Möller, A.; Jakob, G.; Asadi, K. Determining Magnetite/Maghemite Composition and Core−Shell Nanostructure from Magnetization Curve for Iron Oxide Nanoparticles. *J. Phys. Chem. C.* **2018**, *122*, 28292–28301. [CrossRef]
59. Winsett, J.; Moilanen, A.; Paudel, K.; Kamali, S.; Ding, K.; Cribb, W.; Seifu, D.; Neupane, S. Quantitative Determination of Magnetite and Maghemite in Iron Oxide Nanoparticles Using Mössbauer Spectroscopy. *SN Appl. Sci.* **2019**, *1*, 1636. [CrossRef]
60. Singh, A.K.; Srivastava, O.N.; Singh, K. Shape and Size-Dependent Magnetic Properties of Fe_3O_4 Nanoparticles Synthesized Using Piperidine. *Nanoscale Res. Lett.* **2017**, *12*, 298. [CrossRef]
61. LaMer, V.K.; Dinegar, R.H. Theory, Production and Mechanism of Formation of Monodispersed Hydrosols. *J. Am. Chem. Soc.* **1950**, *72*, 4847–4854. [CrossRef]
62. Smith, W.B. Ethylene glycol to acetaldehyde-dehydration or a concerted mechanism. *Tetrahedron* **2002**, *58*, 2091–2094. [CrossRef]
63. Skrabalak, S.E.; Wiley, B.J.; Kim, M.; Formo, E.V.; Xia, Y. On the Polyol Synthesis of Silver Nanostructures: Glycolaldehyde as a Reducing Agent. *Nano Lett.* **2008**, *8*, 2077–2081. [CrossRef] [PubMed]

Article

Sulfidized Nanoscale Zerovalent Iron Supported by Oyster Powder for Efficient Removal of Cr (VI): Characterization, Performance, and Mechanisms

Hao Hu, Donglin Zhao *, Changnian Wu and Rong Xie

Key Laboratory of and Functional Molecule Design and Interface Process, Anhui Jianzhu University, Hefei 230601, China; huhao6031342@163.com (H.H.); wucnustc@126.com (C.W.); xr@ahjzu.edu.cn (R.X.)
* Correspondence: zhaodlin@126.com; Tel.: +86-551-63828100; Fax: +86-551-63828103

Abstract: In this study, sulfidized nanoscale zerovalent iron (S-nZVI) supported by oyster shell (OS) powder (S-nZVI@OS) was synthesized by controlling the initial S/Fe ratios (0.1–0.5) to explore the potential synergistic effects during the adsorption and reduction of Cr (VI). X-ray diffraction (XRD), transmission electron microscopy (TEM), and X-ray photoelectron spectroscopy (XPS) analyses showed that Fe (0) and FeS were well dispersed on the OS surface. Furthermore, the stability of S-nZVI@OS composite was higher than that of nZVI, which was proved by the material ageing experiment. The effects of different S/Fe molar ratios, time, temperature, the initial concentration of Cr (VI), and initial pH on the removal efficiency were also studied. The results indicated that with the increase of the S/Fe molar ratio, the removal capacity of Cr (VI) first increased rapidly and then decreased slowly. Batch experiments showed that an optimal S/Fe molar ratio of 0.2 offered a Cr (VI) removal capacity of about 164.7 mg/g at pH 3.5. The introduction of S can not only promote Cr (VI) reduction but also combine with Cr (III) by forming precipitate on S-nZVI@OS mainly as $Cr_xFe_{(1-x)}OOH$ and Cr_2S_3. The adsorption thermodynamics and kinetics demonstrated that the Langmuir model and pseudo-second-order kinetics model can describe the adsorption isotherms and kinetics. These results suggest that S-nZVI@OS is an effective and safe material for removing Cr (VI) from aqueous solutions.

Keywords: oyster shell; S-nZVI; adsorption; reduction; mechanism

Citation: Hu, H.; Zhao, D.; Wu, C.; Xie, R. Sulfidized Nanoscale Zerovalent Iron Supported by Oyster Powder for Efficient Removal of Cr (VI): Characterization, Performance, and Mechanisms. *Materials* **2022**, *15*, 3898. https://doi.org/10.3390/ma15113898

Academic Editor: Francesco Iacoviello

Received: 15 April 2022
Accepted: 23 May 2022
Published: 30 May 2022

Publisher's Note: MDPI stays neutral with regard to jurisdictional claims in published maps and institutional affiliations.

1. Introduction

Heavy metal–contaminated wastewater is a central problem in water pollution [1,2]. Controlling water pollution and protecting water resources have become essential goals globally. Chromium is one of the primary heavy metals causing groundwater pollution. With rapid industrial development, chromium is widely used in several industries, including electroplating, metallurgy, mechanical engineering, chemical engineering, and electronics [3]. Over the years, with the increase in industrialization worldwide, much chromium-containing wastewater discharge has polluted water supplies [4]. In general, chromium exists in nature in the form of Cr (VI), such as $Cr_2O_7^{2-}$ and CrO_4^{2-} [5–8], and Cr (VI) has been identified as a strong carcinogen [9,10]. The Ministry of Ecology and Environment of China has set the relevant limit values for Cr (VI) [11].

With rises in contamination, many tactics for removing Cr (VI) have been developed, including biological approaches [12], electrochemical oxidation methods, membrane filtering methods [13], and adsorption methods [14]. The adsorption method is regarded as the most promising way of eliminating Cr (VI) due to its high efficiency, cheap cost, and simplicity. The most commonly utilized substance for removing Cr (VI) is nanoscale zerovalent iron (nZVI) [15]. It has a high specific surface area, is highly reducible, and is low in toxicity [16]. However, particle agglomeration reduces the migration ability of nZVI in a porous environment. Moreover, nZVI exhibits low reaction efficiency with other chemicals in waste, which reduces pollutant removal efficiency [17]. The main methods of nZVI modification are emulsification, immobilization, vulcanization, and bimetallic particles. After several years of research, Han and Yang [18] found that nZVI could be modified by sodium thiosulate or sodium sulfide to increase its conductivity, inhibit its reaction with water, and extend its lifespan. Moreover, the nZVI-loaded metal-organic framework exhibited significant Cr (VI) removal capacity [19]. nZVI has been successfully loaded onto silica [20], activated carbon [21,22], zeolite [23], biochar [24], chitosan [25], metal [26], and other materials. However, nZVI materials still have some drawbacks. For example, the process of removing Cr (VI) requires highly acidic conditions, and the reaction produces large amounts of sludge containing heavy metals [27]. To further improve the reducing ability of nZVI, sulphur is usually added to the material to generate S-nZVI, with a stronger reducing ability to remove Cr (VI) [28].

Many sorbents, including biochar [29], clay [30], polymers [31,32] and graphene oxide [33], can be used as carriers. Oyster shell (OS) is a common marine debris in coastal cities. Its chemical constituents are mainly inorganic substances. OS is rich in calcium salts and has a porous surface. The shell comprises a cuticular layer, a prismatic layer and a pearl layer. The cuticular layer is highly resistant to corrosion, the prismatic layer has a foliated structure with a large number of 2~10 microns micropores, and the pearl layer is mainly calcite. Due to its special structure and calcium carbonate composition, OS can be decomposed into CaO and CO during high-temperature calcination, as well as CO_2 gas. Moreover, its pore structure endows it with strong adsorption capacity, exchange capacity, and catalytic decomposition capacity. Therefore, it can absorb various pollutants in sewage and improve water quality. Through dynamic column experiments and static batch experiments, Gao et al. [34–36] studied the adsorption performance of OS powder towards cadmium and cobalt in an aqueous solution. The adsorbent exhibited better Cd^{2+} removal than that of Co^{2+} removal from single-component metal ion solutions. In addition, OS is self-alkaline, which facilitates Cr (III) precipitation. The treatment of domestic waste with OS powder is characterized by high efficiency, low consumption, and the absence of secondary pollution, demonstrating the potential applications of the powder.

It is reported that OS has not been used as a solid loading material for nZVI in the current literature. In this study, OS was used as the loading material to synthesize S-nZVI@OS, which was used to remove Cr (VI) from water. S-nZVI@OS was characterized via XRD, SEM, FTIR, and XPS. On Cr(VI) removal, the influence of pH, Cr(VI) starting concentration, and temperature was examined.

2. Materials and Methods

2.1. Chemical Reagent

Potassium dichromate ($K_2Cr_2O_7$, 99.8%) was purchased from Tianjin Chemical Co., Ltd. Ferrous sulfate heptahydrate ($FeSO_4 \cdot 7H_2O$) and phosphoric acid were purchased from McLaines Biochemical Technology Co., Ltd. (Shanghai, China), and sodium sulfide monohydrate ($Na_2S \cdot 9H_2O$) was bought from Fuchen Reagent Co., Ltd. Sodium borohydride ($NaBH_4$) was purchased from Damao Chemical Reagent Factory (Tianjin, China). Oyster powder was purchased from Guangxi. The original solution containing potassium dichromate (500 mg/L) was prepared by drying 0.2829 g of $K_2Cr_2O_7$ pellets at high temperature for a period of time and then dissolving them in deionized water.

2.2. The Preparation of S-nZVI@OS

Composite diagram of S-nZVI@OS is shown in Figure 1. The S-nZVI@OS was prepared via the following processes: (1) To maximize contact between oyster powder and iron, 1.1 g of oyster powder was dispersed in 5.5 g of 100 mL of $FeSO_4 \cdot 7H_2O$ solution under ultrasonic conditions and stirred for 30 min. (2) The mixture was stirred with a mechanical stirrer under N_2 conditions for 1 h, and then borohydride (3 g, 50 mL) was slowly added with a constant-pressure separator to reduce the iron ions in the solution. (3) $Na_2S \cdot 9H_2O$ was deposited into the mixture. The introduction of sulfide turns the system into FeS, and an FeS layer is formed on the nZVI surface when the system contains a large amount of Fe^{2+}. The resulting material was cleaned with anaerobic water and anhydrous ethanol many times before being vacuum dried and stored in a vacuum glove box until it was needed again.

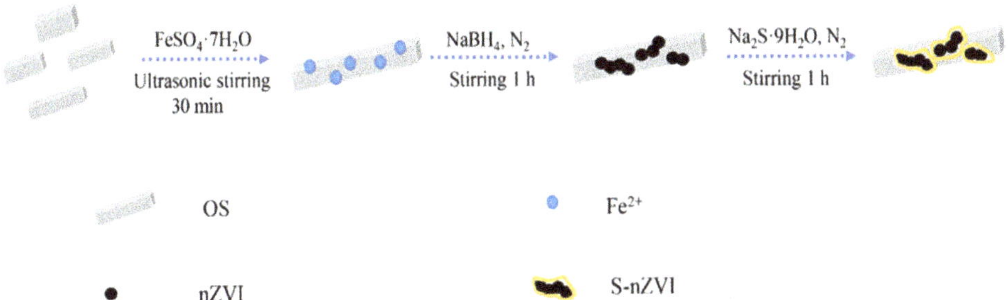

Figure 1. Schematic diagram of S-nZVI@OS synthesis.

2.3. Characterization and Analysis Methods

The composite crystal structures were characterized using X-ray diffraction (XRD, Bruker D8 Advance, Germany) scanned in the range of 10–80° (2θ). The morphology of the product was studied by transmission electron microscope (Jeol Jem-2100, TEM) images and scanning electron microscope (Regulus8100, SEM) images. An electronic spectrometer (XPS, PHI-5300, UK) for analyzing the state of surface elements using X-ray photographs of chemical elements. In addition, a zeta potential meter (Malvern Zetasizer Nano ZS 90, UK) and a vibrating sample magnetometer (VSM LakeShore 7400-S, USA) were used to analyse the surface charge and magnetic properties of the material. Total Cr was determined by (ICP-OES) (Optima 8000, USA) and Cr (VI) concentration was determined by a spectrophotometric UV spectrophotometer (T6 New Century, China).

2.4. Removal Process

In 50 mL test tubes (25 °C), experiments on the adsorption of Cr (VI) by the new material S-nZVI@OS were carried out. A certain concentration of S-nZVI@OS was put into a 20 mg/L solution of Cr (VI) and then the pH was adjusted to 3.5 using various concentrations of HCl or NaOH. Conical flasks were used to study the removed material by shaking them in a temperature-controlled water bath shaker at 180 rpm. At the end of the specified experimental time, the adsorbent material is separated from the water phase using a magnet. All of the experiments were conducted three times.

Equation (1) was used to analyze Cr (VI) removal efficiency (RE), while Equation (2) was used to estimate Cr (VI) removal capacity (RC):

$$RE(\%) = \frac{(C_0 - C_t)}{C_0} \times 100\% \tag{1}$$

$$RC = \frac{(C_0 - C_t) \cdot V}{m} \tag{2}$$

where C_0 is the initial Cr (VI) concentration (mg/L) and C_t is the equilibrium Cr (VI) concentration (mg/L); m is the adsorbent mass (g); and V is the total of the aqueous systems in the reaction system (mL).

Quantitative relationships between Cr (III) concentrations, Cr (VI) concentrations, and total Cr concentrations:

$$C_{Cr3+} + C_{Cr6+} = C_{TCr} \tag{3}$$

where C_{TCr}, C_{Cr6+}, and C_{Cr3+} are the *TCr*, Cr (VI), and Cr (III) concentrations in the reaction system (mg/L), respectively.

2.5. Kinetic Study

Kinetic experiments were performed using a three-necked flask. The experiments were performed at the same temperature under mechanical stirring. The Cr (VI) content was determined according to the amount of supernatant absorbed within a specified time.

The kinetic data were analyzed using the Langmuir–Hinshelwood first-order kinetic model Equation (4) and a pseudo-second-order kinetic model Equation (5) [37,38].

$$\ln(C_t/C_0) = -k_{obs}t + c \tag{4}$$

$$\frac{t}{q_t} = \frac{1}{k_2 q_e^2} + \frac{t}{q_e} \tag{5}$$

where q_e and q_t are the Cr (VI) removal capacities at equilibrium and time t (mg/g), respectively; k_{obs} (min^{-1}) is the Langmuir–Hinshelwood first-order kinetic model's rate constant; c is a constant; and k_2 (g·min/mg) is the pseudo-second-order rate constant.

2.6. Isotherms and Thermodynamics

The Langmuir Equation (6) and Freundlich isotherms Equation (7) are common adsorption models.

$$\text{Langmuir}: \frac{C_e}{q_e} = \frac{1}{q_m b} + \frac{C_e}{q_m} \tag{6}$$

$$\text{Freundlich}: \ln q_e = \ln k + \frac{1}{n} \ln C_e \tag{7}$$

The adsorbent concentration in solution at adsorption equilibrium is C_e (mg/L), q_e (mg/g) is the removal capacity of the material for Cr (VI), q_m is the maximum removal capacity of the material for Cr (VI) (mg/g), and b (L/mg) is the Langmuir model constant associated with the adsorbent to the adsorbate. The Freundlich constant is k, and the adsorption strength is n (L/mg).

The thermodynamic parameters are determined by the following relations:

$$\Delta G^0 = -RT \ln K \tag{8}$$

$$\ln K = \frac{\Delta S}{R} - \frac{\Delta H}{RT} \tag{9}$$

The free energy ΔG^0 (kJ/mol), the enthalpy changes ΔH^0 (kJ/mol), and the entropy change ΔS^0 (J/mol/K) were all calculated. R is the ideal gas constant (8.314 J/(mol·K)) and T is the temperature in Kelvin (K). ln$K$ is obtained by plotting lnK_d as a function of C_e, with C_e extrapolated to zero. ΔS^0 and ΔH^0 are the intercept and slope of a linear plot between ln$K$ and $1/T$, respectively.

3. Results and Discussion

3.1. Characterization

Figure 2 shows SEM and TEM images of the material before and after reaction. SEM image of S-nZVI@OS is shown in Figure 2a. TEM images (Figure 2b,c) show that the dendritic S-nZVI is uniformly dispersed on the surface of the OS carrier and are more dispersed compared to nZVI, so the material has better spatial stability. Figure 2d shows essentially no change in the structure of the material compared to TEM images (Figure 2b,c) before reaction, which is further evidence of the stability of the material.

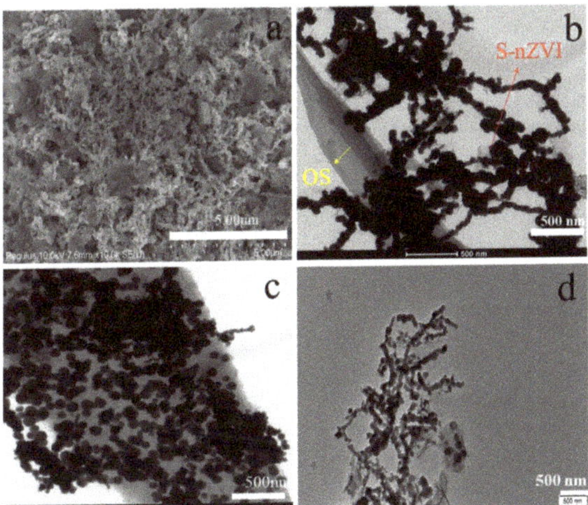

Figure 2. (**a**) SEM image of S-nZVI@OS; (**b,c**) TEM images of S-nZVI@OS; (**d**) TEM image of S-nZVI@OS after reaction.

The elemental composition of S-nZVI@OS (Figure 3b) illustrates the presence of Fe, O and S elements in S-nZVI@OS, demonstrating the formation of Fe oxides and the presence of S after sulphide modification. Figure 3c–h show the HRTEM image of S-nZVI@OS and the corresponding EDS mapping after reaction. In comparison to Figure 3b, the mass concentration of S before reaction reduced from 0.57% to 0.16% after reaction, whereas the mass content of Fe reduced from 88.53% to 78.73%. This indicates that oxidation of S and Fe occurs during the removal process, resulting in the formation of SO_4^{2-} and oxygenated iron compounds.

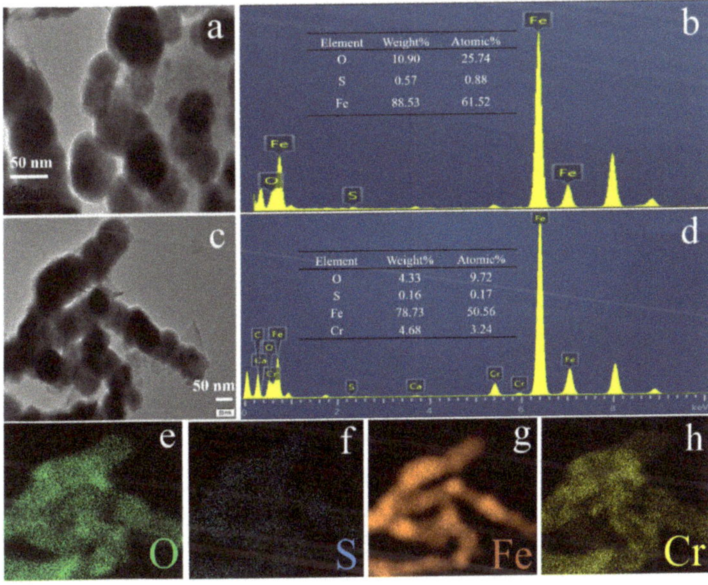

Figure 3. (**a,b**) TEM image of S-nZVI@OS and corresponding EDS images before reaction; (**c–h**) TEM image of S-nZVI@OS and corresponding EDS mapping images after reaction.

Figure 4a describes powder XRD patterns of different composite before and after removal Cr (VI). The peak at 2θ of 29.4° is $CaCO_3$ (PDF#05-0586), corresponding to the (104) plane of the index. In the S-nZVI@OS composite, the crystalline metals Fe (PDF#99-0064) and FeS (PDF#23-1123) were present due to the diffraction peaks at 44.67° and 47.46°, corresponding to the index (110) and (220) planes, respectively. After reaction with Cr (VI), new characteristic peaks appeared at 35.62° and 62.59° indexed as (3 1 1) and (2 1 4) planes corresponding to the characteristic peaks of Fe_3O_4 (PDF#88-0315) and Fe_2O_3 (PDF#72-0469). The presence of iron oxides after reaction indicates that the iron was oxidized during the removal process. The XRD pattern of OS after reaction with Cr (VI) shows there is no characteristic peak of new substance, which indicates that OS mainly participates in the adsorption.

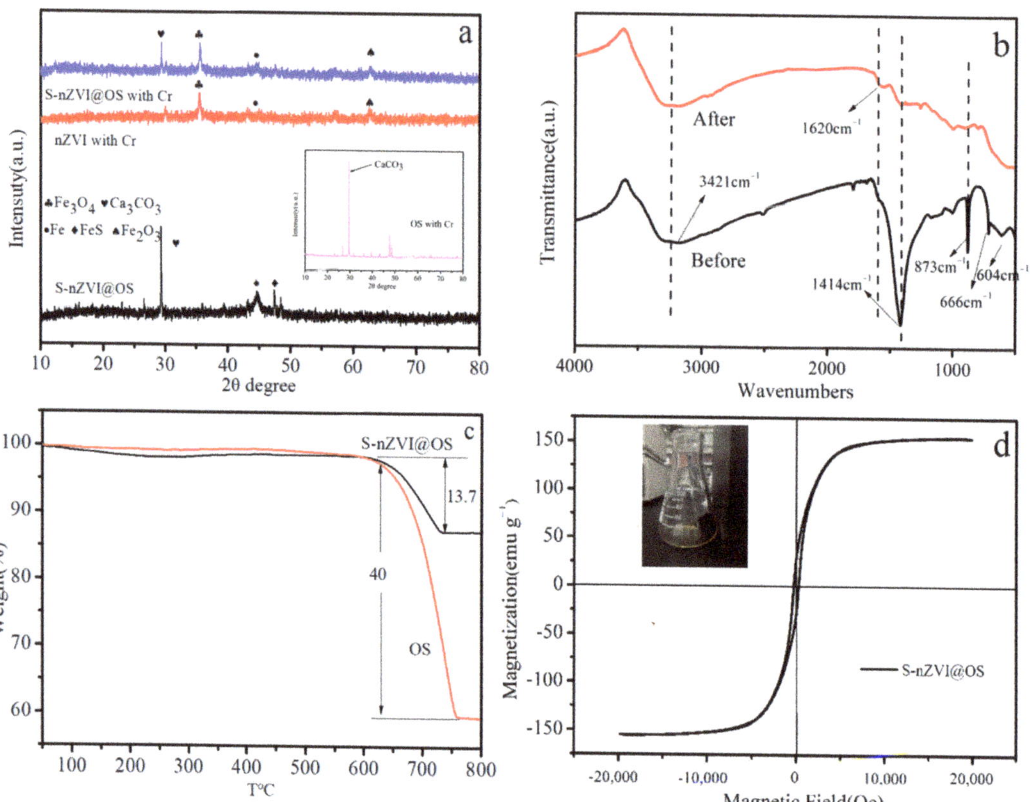

Figure 4. (**a**) XRD patterns of different composite before and after removal Cr (VI) (inset figure is the XRD pattern of OS after reaction with Cr (VI)); (**b**) FTIR spectra of S-nZVI@OS before and after reaction; (**c**) TG thermogram of OS and S-nZVI@OS; (**d**) magnetization curve of S-nZVI@OS.

The FTIR spectra of the composite material before and after the reaction are shown in Figure 4b. Prior to the reaction, peak locations are 3421 cm⁻¹, 1620 cm⁻¹, 1414 cm⁻¹, and 873 cm⁻¹, corresponding to stretching vibrations related to hydroxyl (-OH), C=O bonds, carboxyl (-COOH), and C-O bonds, respectively. The disappearance of absorption vibration absorption peaks for carboxyl and C-O bonds following reaction indicates a reduction reaction during Cr (VI) removal. In addition, the spectrum of the S-nZVI@OS composite exhibits absorption peaks at 666 and 604 cm⁻¹, corresponding to the formation of Fe-O and Fe-S bonds during the preparation of the composite [39]. In contrast, the vibrational absorption peak here disappears after the reaction, indicating that the oxidation of Fe and S occurred during the removal process.

Figure 4c shows the results of the thermogravimetric analysis of OS and S-nZVI@OS. Both materials have mass loss at 630 °C, and the weight loss of OS is 40%, which is due to the decomposition of $CaCO_3$ contained in OS. S-nZVI@OS loses 13.7% of its weight owing to the decomposition of OS, which indicates that S-nZVI loaded by OS has good thermal stability. Figure 4d is the magnetization curve of S-nZVI@OS. It can be clearly seen that the magnetization value of S-nZVI@OS is high (97 emu/g), which indicates that the adsorbent can be recycled using magnetism well after the reaction is completed.

3.2. Adsorption Kinetics for S-nZVI@OS with Different S/Fe Ratio

Figure 5a depicts the kinetics of Cr (VI) elimination for materials with various S/Fe ratios. The Cr (VI) removal rate was high for the first 30 min and then levelled off with time. The equilibrium adsorption capacities were 100, 158, 125, and 112 mg/g for S/Fe ratios of 0.1, 0.2, 0.35, and 0.5, respectively. The results reveal that as the S/Fe ratio increases, the Cr (VI) removal capability increases at first, then falls. A too-high S/Fe ratio may lead to the formation of less active FeS_n, which reduces the reduction capacity of Fe (0) and the activity of the particles [40]. Therefore, the S/Fe ratio affects the effectiveness of the material in removing Cr (VI).

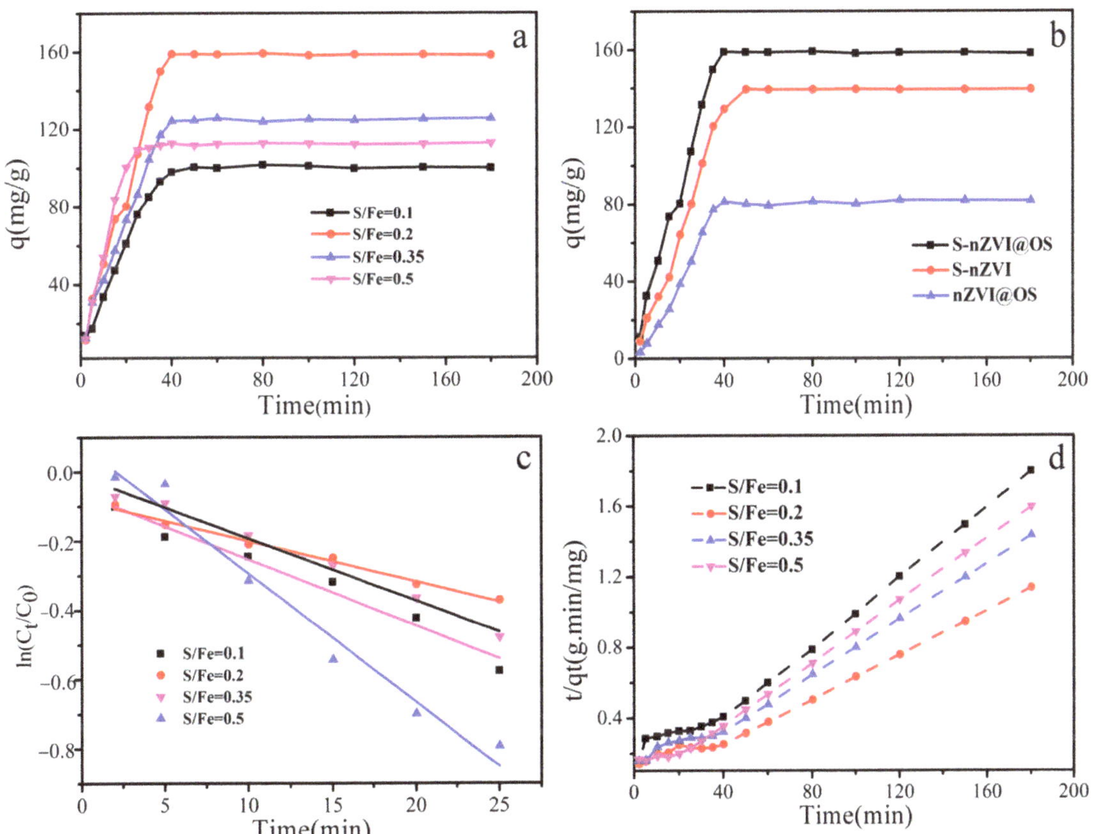

Figure 5. (a) The adsorption kinetics of Cr (VI) with different S/Fe ratios; (b) the adsorption kinetics of S-nZVI, nZVI@OS and S-nZVI@OS; (c) Liner polts of lnC_t/C_0 versus T; (d) pseudo-second-order dynamics diagram.

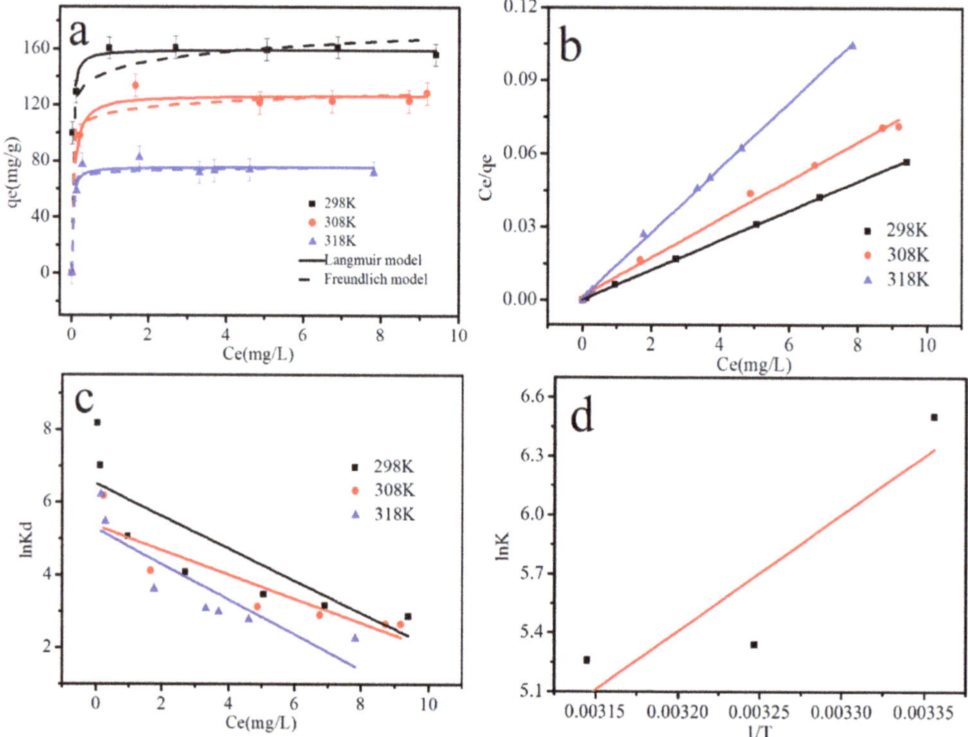

Figure 7. (**a**) Adsorption isotherms for Cr (VI) on S-nZVI@OS at 298, 308 and 318 K; (**b**) Langmuir plot; (**c**) linear plots of $\ln K_d$ versus C_e; (**d**) linearized Arrhenius plot of $\ln K$ as a function of $1/T$.

Table 2. Isotherm parameters of S-nZVI@OS removal of Cr (VI).

T(K)	Langmuir Model			Freundlich Model		
	Q_{max} (mg/g)	b (L/mg)	R^2	k (mg·g^{-1})	n	R^2
298	164.745	21.328	0.999	143.314	12.402	0.908
308	126.582	4.463	0.994	104.921	14.760	0.760
318	75.075	11.684	0.998	66.567	17.077	0.925

Table 3. Comparison of Cr (VI) removal by various modified nZVI materials.

Adsorbents	pH	Q_{max} (mg/g)	References
S-nZVI@OS	3.5	164.7	This work
Biochar-CMC-nZVI	5.6	112.5	[45]
nZVI/Cu	5.0	18.8	[46]
nZVI@HCl-BC	5.0	17.8	[47]
TP-nZVI-OB	2.0	95.5	[48]
SBC-nZVI	3.0	84.4	[49]
nGO-nZVI	7.0	21.7	[50]
Sepiolite/nZVI	6.0	43.9	[51]
CS-nZVI	4.0	101.8	[52]

Table 4. Thermodynamic parameters of Cr (VI) removal by S-nZVI@OS at different temperatures (298, 308, 318 K).

T (K)	ΔG^0 (kJ/mol)	ΔH^0 (kJ/mol)	ΔS^0 (J/mol/K)
298	−15.695		
308	−14.562	−49.457	−113.294
318	−13.429		

3.5. Effect of Material Ageing on Removal Efficiency

Agglomeration and anodization of the nZVI layer occur with age, reducing the surface activity of nZVI and affecting Cr (VI) removal [53]. In the current study, the S-nZVI@OS composite maintained a high Cr (VI) removal rate after 14 days (Figure 8). FeS_X possibly reduced the magnetic attraction between nZVI particles, thereby inhibiting particle agglomeration. Moreover, the sulfide will inhibit the side reaction of the material with water, which helps to preserve the high activity of the material. Comparing the TEM images before and after the reaction, the structure of the material is almost unchanged before and after the reaction. This further demonstrates the stability of the new composite S-nZVI@OS in terms of Cr (VI) removal [54].

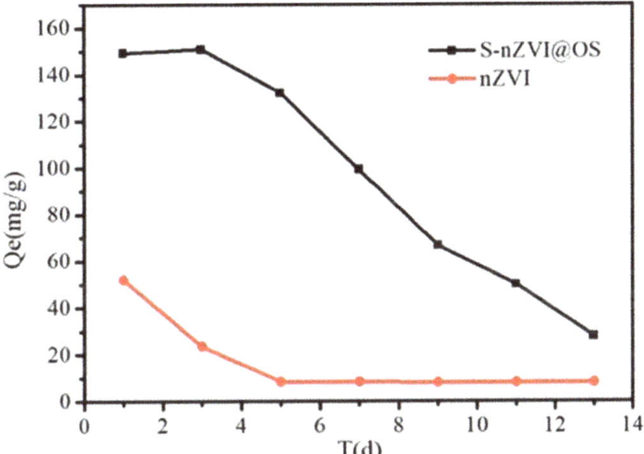

Figure 8. Changes in adsorption capacity of S-nZVI@OS and nZVI@OS over time.

4. Mechanism Analysis

The Fe 2p3/2 XPS spectra of S-nZVI@OS are shown in Figure 9b. Fe^0 and Fe (III) are represented by peaks with binding energies of 707.5 eV and 724.3 eV, respectively, whereas Fe^{2+} is represented by peaks with binding energies of 711.6 eV and 721.9 eV [53]. The spectrum of the fresh S-nZVI@OS (i.e., before reaction) exhibited the Fe^0 characteristic peak. The peaks of Fe^0 and Fe^{2+} move to high binding energy peak positions in the spectrum of S-nZVI@OS (i.e., after reaction). The peaks at 724.3 eV and 725 eV represent the production of FeOOH and $FeCr_2O_4$. This suggests that the process occurs with the oxidation of Fe^0 and the precipitation formation of FeOOH.

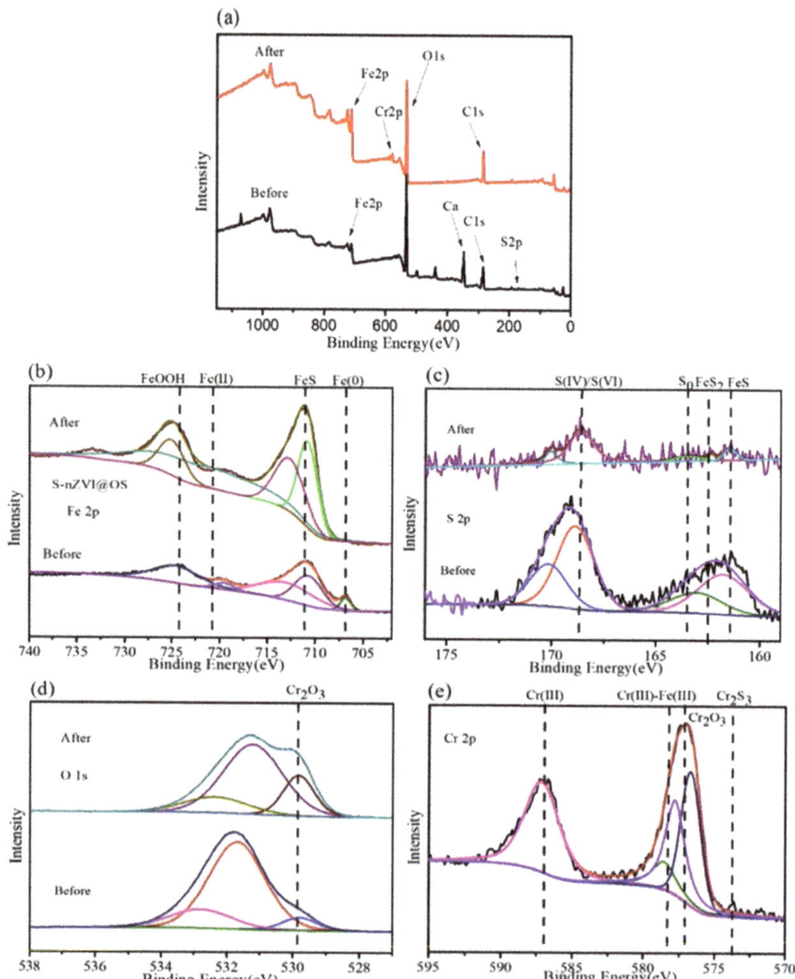

Figure 9. (**a**) XPS survey of S-nZVI@OS before and after reaction with Cr (VI); (**b**) Fe2p spectra before and after reaction; (**c**) S2p spectra before and after reaction; (**d**) O1s spectra before and after reaction; (**e**) Cr2p spectrum.

The S 2p spectrum (Figure 9c) prior to the reaction shows peaks at 161.9, 164.3, 162.5, and 169 eV, corresponding to S (-II), S (0), FeS_2, and S (IV), respectively [55]. From the fractional spectrum of S 2p before and after the reaction (Figure 9c), S is mainly present as S (-II), FeS_2, S (IV), and S (VI). Thus, S (-II) and S (0) were oxidized to S (IV)/S (VI) by Cr (VI) under acidic conditions (pH = 3.5). This indicates that Cr removal was also based on a reduction reaction.

The O 1s spectrum of S-nZVI@OS (Figure 9d) exhibited three peaks. Three peaks appear at 530.1, 531.6 and 533.2 eV, which correspond to an oxide (O^{2-}), a surface hydroxyl group (-OH) and adsorbed water (H_2O), respectively. Iron oxide and hydroxide compounds existed on the S-nZVI@OS surface before the reaction. After the reaction, the intensity of the oxide (O^{2-}) peak increased, and it can be judged that the main product was Cr_2O_3.

Figure 9e shows the Cr 2p XPS spectrum. Three peaks with binding affinity of 578.7, 587, and 574 eV occur, with 578.7 eV corresponding to Cr (III)-Fe (III) and 587 and 574 eV belonging to Cr (III) and Cr_2S_3, respectively [56]. The Cr (VI) binding energy weakens after the reaction, whereas the Cr (III)-Fe (III) and Cr (III) binding energies become significantly stronger. These modifications suggest that Cr (VI) was reduced and forms Cr (III)-Fe (III) complexes on the adsorbent surface.

The removal mechanism is shown in Figure 10. The elimination of Cr (VI) can be separated into three steps, according to the analysis. Firstly, Cr (VI) can be trapped just on material surface due to its substantial positive charge and porosity. Secondly, electrons are transferred from Fe^0, Fe^{2+}, S^{2-}, and S to Cr (VI), which is reduced to Cr (III). Fe^0, Fe^{2+} and S are oxidized to Fe_2O_3, Fe_3O_4, and SO_4^{2-}. Finally, the metal oxides contained in OS will generate a large number of metal cations (Ca^{2+}) under acidic conditions. Cations in the aqueous solution are hydrolyzed to generate a large amount of OH⁻, thus generating $(Cr_xFe_{(1-x)})$ $(OH)_3$ and $Cr_xFe_{(1-x)}$ OOH precipitates. A considerable amount of H^+ in the solution is consumed during the electron transfer process. Therefore, the solution pH increased from 3.5 to 5.8. In summary, the OS played an essential role in not only the dispersion, maintenance, and minimization of S-nZVI particles but also the adsorption and precipitation processes. The specific reaction process was presented as Equations (10)–(15) [56,57]. The combined action of Fe (0), S (0), Fe^{2+}, and S^{2-} removed Cr (VI). The reaction requires fewer hydrogen ions than those associated with other materials, which makes S-nZVI@OS advantageous. Furthermore, oxidized S, Fe (III), and Cr (III) were formed as sulfates and precipitates, reducing the material's passivation effect.

$$Cr_2O_7^{2-} + 3Fe^0 + 14H^+ \rightarrow 2Cr^{3+} + 3Fe^{2+} + 7H_2O \tag{10}$$

$$(1-x)Fe^{3+} + xCr^{3+} + 3H_2O \rightarrow \left(Cr_xFe_{(1-x)}\right)(OH)_3(s) + 3H^+ \tag{11}$$

$$(1-x)Fe^{3+} + xCr^{3+} + 2H_2O \rightarrow Cr_xFe_{(1-x)}OOH(s) + 3H^+ \tag{12}$$

$$HCrO_4^- + 3Fe^{2+} + 7H^+ \rightarrow Cr^{3+} + 3Fe^{3+} + 4H_2O \tag{13}$$

$$Cr_2O_7^{2-} + S^0 + 5H^+ \rightarrow 2Cr^{3+} + SO_4^{2-} + 3H_2O \tag{14}$$

$$2Cr^{3+} + 3S^{2-} \rightarrow Cr_2S_3(s) \tag{15}$$

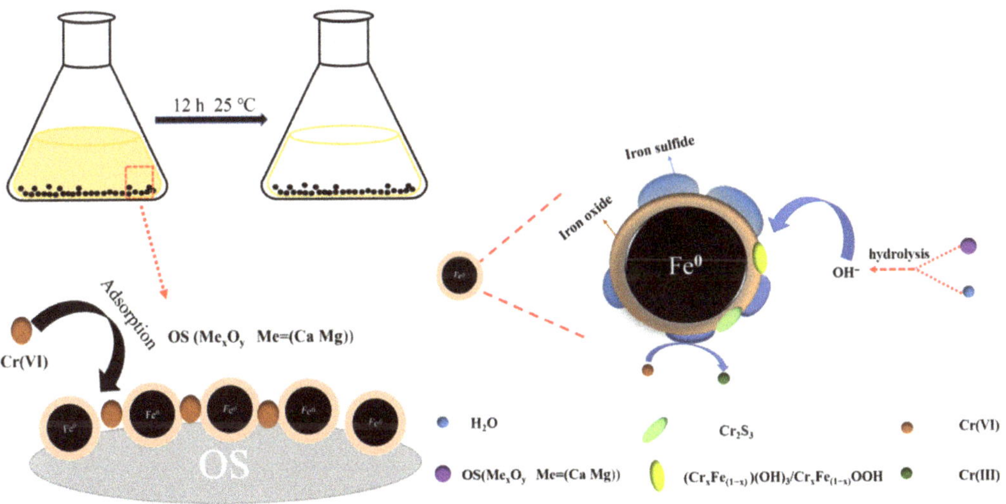

Figure 10. Schematic representation for mechanistic pathway of Cr (VI) reduction by S-nZVI@OS.

5. Conclusions

S-nZVI@OS, a ternary composite prepared in two steps, successfully removed Cr (VI) from water. TEM and SEM results demonstrated that the particles were uniformly dispersed, with minimal agglomeration. The Langmuir isotherm model well describes the experimental data. The Langmuir–Hinshelwood first-order kinetic model and a pseudo-second-order kinetic model agreed well with the kinetics of Cr (VI) removal. Under acidic conditions (low solution pH), the new ternary composite S-nZVI@OS exhibits a powerful removal of Cr (VI). In addition, S-nZVI@OS has shown excellent performance in terms of stability. The highest removal rate was achieved at a solution pH of 3.5 with an S-nZVI@OS dosage of 0.1 g/L. Adsorption isotherms were obtained at 298, 308 and 318 K. The free energies ΔG^0 (kJ/mol) were obtained as -15.695, -14.562 and -13.429 kJ/mol, respectively, the enthalpy changes ΔH^0 (kJ/mol) and entropy change ΔS^0 (J/mol/K) were -49.457 kJ/mol and -113.294 J/mol/K. The results showed that the lower temperature was favorable for the reaction, and the corresponding adsorption capacity of S-nZVI@OS was 164.7 mg/g at 298 K. XPS analysis shows that the process of S-nZVI@OS in the removal of Cr (VI) is an integrated adsorption and reduction process. Self-alkaline loading material OS not only has excellent adsorption performance, but also contributes to the formation of $(Cr_xFe_{(1-x)})(OH)_3$ and $Cr_xFe_{(1-x)}OOH$ precipitates through hydrolysis. To summarize, OS-supported S-nZVI@OS is a low-cost, effective, and ecologically friendly Cr (VI) removal material.

Author Contributions: H.H.: investigation, formal analysis, visualization, writing-review and editing. D.Z.: project administration, funding acquisition. C.W.: data curation, project administration R.X.: formal analysis, project administration. All authors have read and agreed to the published version of the manuscript.

Funding: Financial support from the National Natural Science Foundation of China (21876001), Special Support Plan of Anhui Province and the University Synergy Innovation Program of Anhui Province (GXXT-2021-012) are acknowledged.

Institutional Review Board Statement: Not applicable.

Informed Consent Statement: Not applicable.

Data Availability Statement: The data presented in this study are available on request from the corresponding author.

Conflicts of Interest: The authors declare no conflict of interest.

References

1. Dhal, B.; Thatoi, H.N.; Das, N.N.; Pandey, B.D. Chemical and microbial remediation of hexavalent chromium from contaminated soil and mining/metallurgical solid waste: A review. *J. Hazard. Mater.* **2013**, *250–251*, 272–291. [CrossRef] [PubMed]
2. Li, J.; Wang, X.X.; Zhao, G.X.; Chen, C.L.; Chai, Z.F.; Alsaedi, A.; Hayat, T.; Wang, X.K. Metal-organic framework-based materials: Superior adsorbents for the capture of toxic and radioactive metal ions. *Chem. Soc. Rev.* **2018**, *47*, 2322–2356. [CrossRef] [PubMed]
3. Zhu, K.R.; Chen, C.L.; Lu, S.H.; Zhang, X.D.; Alsaedi, A.; Hayat, T. MOFs-induced encapsulation of ultrafine Ni nanoparticles into 3D N-doped graphene-CNT frameworks as a recyclable catalyst for Cr (VI) reduction with formic acid. *Carbon* **2019**, *148*, 52–63. [CrossRef]
4. Li, S.L.; Wang, W.; Liu, Y.Y.; Zhang, W.X. Zero-valent iron nanoparticles (nZVI) for the treatment of smelting wastewater: A pilot-scale demonstration. *Chem. Eng. J.* **2014**, *254*, 115–123. [CrossRef]
5. Sarin, V.; Singh, T.S.; Pant, K.K. Thermodynamic and breakthrough column studies for the selective sorption of chromium from industrial effluent on activated eucalyptus bark. *Bioresour. Technol.* **2006**, *97*, 1986–1993. [CrossRef]
6. Wang, J.Y.; Li, Y.C.; Xie, Y.; Zhu, K.R.; Alsaedi, A.; Hayat, T.; Chen, C.L. Construction of novel graphene-based materials GO@SiO$_2$@C@Ni for Cr (VI) removal from aqueous solution. *J. Colloid Interface Sci.* **2019**, *557*, 254–269. [CrossRef]
7. Bhaumik, M.; Maity, A.; Srinivasu, V.V.; Onyango, M.S. Enhanced removal of Cr (VI) from aqueous solution using polypyrrole/Fe$_3$O$_4$ magnetic nanocomposite. *J. Hazard. Mater.* **2011**, *190*, 381–390. [CrossRef]
8. Gong, K.D.; Hu, Q.; Xiao, Y.Y.; Cheng, X.; Liu, H.; Wang, N.; Qiu, B.; Guo, Z.H. Triple layered core-shell ZVI@carbon@polyaniline composite enhanced electron utilization in Cr (VI) reduction. *J. Mater. Chem.* **2018**, *A6*, 11119–11128. [CrossRef]
9. Lei, C.S.; Zhu, X.F.; Zhu, B.C.; Jiang, C.J.; Le, Y.; Yu, J.G. Superb adsorption capacity of hierarchical calcined Ni/Mg/Al layered double hydroxides for Congo red and Cr(VI) ions. *J. Hazard. Mater.* **2017**, *321*, 801–811. [CrossRef]

10. Tan, Y.; Zou, Q.; Liu, X.F. Adsorption behavior comparison of trivalent and hexavalent chromium on biochar derived from municipal sludge. *Bioresour. Technol.* **2015**, *190*, 388–394. [CrossRef]

11. Liu, C.; Fiol, N.; Villaescusa, I.; Poch, J. New approach in modeling Cr (VI) sorption onto biomass from metal binary mixtures solutions. *Sci. Total Environ.* **2016**, *541*, 101–108. [CrossRef] [PubMed]

12. Rengaraj, S.; Joo, C.K.; Kim, Y.; Yi, J. Kinetics of removal of chromium from water and electronic process wastewater by ion exchange resins: 1200H, 1500H and IRN97H. *J. Hazard. Mater.* **2003**, *102*, 257–275. [CrossRef]

13. Kaya, A.; Onac, C.; Alpoguz, H.K.; Yilmaz, A.; Atar, N. Removal of Cr(VI) through calixarene based polymer inclusion membrane from chrome plating bath water. *Chem. Eng. J.* **2016**, *283*, 141–149. [CrossRef]

14. Putz, A.; Ciopec, M.; Negrea, A.; Grad, O.; Ianăşi, C.; Ivankov, O.; Milanoviʼc, M.; Stijepoviʼc, I.; Almásy, L. Comparison of Structure and Adsorption Properties of Mesoporous Silica Functionalized with Aminopropyl Groups by the Co-Condensation and the Post Grafting Methods. *Materials* **2021**, *14*, 628. [CrossRef]

15. Stefaniuk, M.; Oleszczuk, P.; Yong, S.O. Review on nano zerovalent iron (nZVI): From synthesis to environmental applications. *Chem. Eng. J.* **2016**, *287*, 618–632. [CrossRef]

16. Fu, F.; Dionysiou, D.D.; Liu, H. The use of zero-valent iron for groundwater remediation and wastewater treatment: A review. *J. Hazard. Mater.* **2014**, *267C*, 194–205. [CrossRef]

17. Wan, Z.H.; Chu, D.W.; Tsang, D.C.W.; Li, M.; Sun, T.; Verpoort, F. Concurrent adsorption and micro-electrolysis of Cr (VI) by nanoscale zerovalent iron/biochar/Ca-alginate composite. *Environ. Pollut.* **2019**, *247*, 410–420. [CrossRef]

18. Han, Y.; Yan, W. Reductive dechlorination of trichloroethene by zero-valent iron nanoparticles: Reactivity enhancement through sulfidation treatment. *Environ. Sci. Technol.* **2016**, *50*, 2992–13001. [CrossRef]

19. Fang, Y.; Wen, J.; Zhang, H.; Wang, Q.; Hu, X. Enhancing Cr (VI) reduction and immobilization by magnetic core-shell structured NZVI@MOF derivative hybrids. *Environ. Pollut.* **2020**, *260*, 114021. [CrossRef]

20. Fang, W.; Jiang, X.Y.; Luo, H.J.; Geng, J.J. Synthesis of graphene/SiO$_2$@polypyrrole nanocomposites and their application for Cr (VI) removal in aqueous solution. *Chemosphere* **2018**, *197*, 594–602. [CrossRef]

21. Wang, Y.H.; Lin, S.H.; Juang, R.S. Removal of heavy metal ions from aqueous solutions using various low-cost adsorbents. *J. Hazard. Mater.* **2003**, *102*, 291–302. [CrossRef]

22. Al-Othman, Z.A.; Ali, R.; Naushad, M. Hexavalent chromium removal from aqueous medium by activated carbon prepared from peanut shell: Adsorption kinetics, equilibrium and thermodynamic studies. *Chem. Eng. J.* **2012**, *184*, 238–247. [CrossRef]

23. Yusof, A.M.; Malek, N.A.N.N. Removal of Cr(VI) and As(V) from aqueous solutions by HDTMA-modified zeolite Y. *J. Hazard. Mater.* **2009**, *162*, 1019–1024. [CrossRef]

24. Dong, X.L.; Ma, L.Q.; Li, Y.C. Characteristics and mechanisms of hexavalent chromium removal by biochar from sugar beet tailing. *J. Hazard. Mater.* **2011**, *190*, 909–915. [CrossRef] [PubMed]

25. Kyzas, G.Z.; Siafaka, P.I.; Pavlidou, E.G.; Chrissafis, K.J.; Bikiaris, D.N. Synthesis and adsorption application of succinyl-grafted chitosan for the simultaneous removal of zinc and cationic dye from binary hazardous mixtures. *Chem. Eng. J.* **2015**, *259*, 438–448. [CrossRef]

26. Hu, L.Y.; Chen, L.X.; Liu, M.T.; Wu, L.J.; Feng, J.J. Theophylline-assisted, eco-friendly synthesis of PtAu nanospheres at reduced graphene oxide with enhanced catalytic activity towards Cr(VI) reduction. *J. Colloid Interface Sci.* **2017**, *493*, 94–102. [CrossRef] [PubMed]

27. Lei, C.; Sun, Y.; Tsang, D.C.W.; Lin, D. Environmental transformations and ecological effects of iron-based nanoparticles. *Environ. Pollut.* **2018**, *232*, 10–30. [CrossRef]

28. Gao, J.; Yang, L.; Liu, Y.; Shao, F.; Liao, Q.; Shang, J. Scavenging of Cr (VI) from aqueous solutions by sulfide-modified nanoscale zero-valent iron supported by biochar. *J. Taiwan Inst.* **2018**, *91*, 449–456. [CrossRef]

29. Qian, L.B.; Zhang, W.Y.; Yan, J.C.; Han, L.; Chen, Y.; Ouyang, D.; Chen, M.F. Nanoscale zero-valent iron supported by biochars produced at different temperatures: Synthesis mechanism and effect on Cr (VI) removal. *Environ. Pollut.* **2017**, *223*, 153–160. [CrossRef]

30. Zhang, Y.L.; Li, Y.M.; Li, J.F.; Sheng, G.D.; Zhang, Y.; Zheng, X.M. Enhanced Cr (VI) removal by using the mixture of pillared bentonite and zero-valent iron. *Chem. Eng. J.* **2012**, *185–186*, 243–249. [CrossRef]

31. Bhaumik, M.; McCrindle, R.I.; Maity, A. Enhanced adsorptive degradation of Congo red in aqueous solutions using polyaniline/Fe^0 composite nanofibers. *Chem. Eng. J.* **2015**, *260*, 716–729. [CrossRef]

32. Huang, R.F.; Ma, X.G.; Li, X.; Guo, L.H.; Xie, X.W.; Zhang, M.Y.; Li, J. A novel ion imprinted polymer based on graphene oxide-mesoporous silica nanosheet for fast and efficient removal of chromium (VI) from aqueous solution. *J. Colloid Interface Sci.* **2018**, *514*, 544–553. [CrossRef] [PubMed]

33. Li, J.; Chen, C.L.; Zhang, R.; Wang, X.K. Nanoscale zero-valent iron particles supported on reduced graphene oxides by using a plasma technique and their application for removal of heavy-metal ions. *Chem. Asian J.* **2015**, *10*, 1410–1417. [CrossRef] [PubMed]

34. Lee, H.H.; Kim, S.Y.; Owens, V.N.; Park, S.; Kim, J.; Hong, C.O. How does oyster shell immobilize cadmium? *Arch. Environ. Contam. Toxicol.* **2018**, *74*, 114–120. [CrossRef]

35. Ok, Y.S.; Lim, J.E.; Moon, D.H. Stabilization of Pb and Cd contaminated soils and soil quality improvements using waste oyster shells. *Environ. Geochem. Health* **2011**, *33*, 83–91. [CrossRef]

36. Xu, X.; Liu, X.; Oh, M.; Park, J. Oyster shell as a low-cost adsorbent for removing heavy metal ions from wastewater. *Pol. J. Environ. Stud.* **2019**, *28*, 2949–2959. [CrossRef]

37. Zhao, D.L.; Gao, X.; Chen, S.H.; Xie, F.Z.; Feng, S.J.; Alsaedi, A.; Hayat, T.; Chen, C.L. Interaction between U(VI) with sulfhydryl groups functionalized graphene oxides investigated by batch and spectroscopic techniques. *J. Colloid Interface Sci.* **2018**, *524*, 129–138. [CrossRef]

38. Zhang, Q.; Zhao, D.L.; Feng, S.J.; Wang, Y.Y.; Jin, J.; Alsaedi, A.; Hayat, T.; Chen, C.L. Synthesis of nanoscale zero-valent iron loaded chitosan for synergistically enhanced removal of U(VI) based on adsorption and reduction. *J. Colloid Interface Sci.* **2019**, *552*, 735–743. [CrossRef]

39. Singh, P.; Raizada, P.; Kunari, S. Solar-Fenton removal of malachite green with novelFe-0-activated carbon nanocomposite. *Appl. Catal. A Gen.* **2014**, *476*, 9–18. [CrossRef]

40. Luo, S.L.; Xu, X.L.; Zhou, G.Y. Amino siloxane oligomer-linked graphene oxide as an efficient adsorbent for removal of Pb(II) from wastewater. *J. Hazard. Mater.* **2014**, *274*, 145–155. [CrossRef]

41. Wang, H.; Wang, X.; Ma, J.; Xia, P.; Zhao, J. Removal of cadmium (II) from aqueous solution: A comparative study of raw attapulgite clay and a reusable waste-struvite/attapulgite obtained from nutrient-rich wastewater. *J. Hazard. Mater.* **2017**, *329*, 66–76. [CrossRef] [PubMed]

42. Yang, Y.; Wang, G.; Wang, B.; Li, Z.; Jia, X.; Zhou, Q. Biosorption of Acid Black 172 and Congo Red from aqueous solution by nonviable Penicillium YW 01: Kinetic study, equilibrium isotherm and artificial neural network modeling. *Bioresour. Technol.* **2011**, *102*, 828–834. [CrossRef] [PubMed]

43. Lyu, H.; Gong, Y.; Tang, J.; Huang, Y.; Wang, Q. Immobilization of heavy metals in electroplating sludge by biochar and iron sulfide. *Environ. Sci. Pollut. Res. Int.* **2016**, *23*, 14472–14488. [CrossRef]

44. Ai, Y.; Liu, Y.; Lan, W.; Jin, J.; Xing, J.; Zou, Y.; Zhao, C.; Wang, X. The effect of pH on the U(VI) sorption on graphene oxide (GO): A theoretical study. *Chem. Eng. J.* **2018**, *343*, 460–466. [CrossRef]

45. Zhang, S.; Lyu, H.H.; Tang, J.C.; Song, B.; Zhen, M.; Liu., X. A novel biochar supported CMC stabilized nano zero-valent iron composite for hexavalent chromium removal from water. *Chemosphere* **2019**, *217*, 686–694. [CrossRef]

46. Zhu, F.; Ma, S.Y.; Liu, T.; Deng, X.Q. Green synthesis of nano zero-valent iron/Cu by green tea to remove hexavalent chromium from groundwater. *J. Clean. Prod.* **2017**, *174*, 184–190. [CrossRef]

47. Dong, H.R.; Deng, J.M.; Xie, Y.K.; Zhang, C.; Jiang, Z.; Cheng, Y.J.; Hou, K.J.; Zeng, G.G. Stabilization of nanoscale zero-valent iron (nZVI) with modified biochar for Cr(VI) removal from aqueous solution. *J. Hazard. Mater.* **2017**, *332*, 79–86. [CrossRef]

48. Zhang, Y.T.; Jiao, X.Q.; Liu, N.; Lv, J.; Yang, Y.D. Enhanced removal of aqueous Cr(VI) by a green synthesized nanoscale zero-valent iron supported on oak wood biochar. *Chemosphere* **2020**, *245*, 125542. [CrossRef]

49. Wu, B.; Peng, D.H.; Hou, S.Y.; Tang, B.C.; Wang, C.; Xu, H. Dynamic study of Cr(VI) removal performance and mechanism from water using multilayer material coated nanoscale zerovalent iron. *Environ. Pollut.* **2018**, *240*, 717–724. [CrossRef]

50. Li, X.; Ai, L.; Jing, J. Nanoscale zerovalent iron decorated on graphene nanosheets for Cr(VI) removal from aqueous solution: Surface corrosion retard induced the enhanced performance. *Chem. Eng. J.* **2016**, *288*, 789–797. [CrossRef]

51. Fu, R.B.; Yang, Y.P.; Xu, Z.; Zhang, X.; Guo, X.P.; Bi, D.S. The removal of chromium (VI) and lead (II) from groundwater using sepiolite-supported nanoscale zero-valent iron (S-NZVI). *Chemosphere* **2015**, *138*, 726–734. [CrossRef] [PubMed]

52. Fan, H.; Ren, H.; Ma, X.; Zhou, S.; Huang, J.; Jiao, W.; Qi, G.; Liu, Y. High-gravity continuous preparation of chitosan-stabilized nanoscale zero-valent iron towards Cr (VI) removal. *Chem. Eng. J.* **2020**, *390*, 124639. [CrossRef]

53. Su, Y.M.; Adeleye, Y.M.; Keller, A.A.; Huang, Y.X.; Dai, C.M.; Zhou, X.F.; Zhang, Y.L. Magnetic sulfide-modified nanoscale zerovalent iron (S-nZVI) for dissolved metal ion removal. *Water Res.* **2015**, *74*, 47–57. [CrossRef] [PubMed]

54. Zhao, R.R.; Zhou, Z.M.; Zhao, X.D.; Jing, G.H. Enhanced Cr(VI) removal from simulated electroplating rinse wastewater by amino-functionalized vermiculite-supported nanoscale zero-valent iron. *Chemosphere* **2019**, *218*, 458–467. [CrossRef]

55. Fontecha-Cámara, M.A.; Álvarez-Merino, M.A.; Carrasco-Marín, F.; Lopez-Ramon, M.V.; Moreno-Castilla, C. Heterogeneous and homogeneous Fenton processes using activated carbon for the removal of the herbicide amitrole from water. *Appl. Catal. B Environ.* **2011**, *101*, 425–430. [CrossRef]

56. Montesinos, V.N.; Quici, N.; Halac, E.B.; Leyva, A.G.; Custo, G.; Bengio, S. Highly efficient removal of Cr (VI) from water with nanoparticulated zerovalent iron: Understanding the Fe(III)-Cr(III) passive outer layer structure. *Chem. Eng. J.* **2014**, *244*, 569–575. [CrossRef]

57. Lyu, H.H.; Tang, J.C.; Huang, Y.; Gai, L.S.; Zeng, E.Y.; Liber, K.; Gong, Y.Y. Removal of hexavalent chromium from aqueous solutions by a novel biochar supported nanoscale iron sulfide composite. *Chem. Eng. J.* **2017**, *322*, 516–524. [CrossRef]

Article

Effect of Polyimide-Phosphating Double Coating and Annealing on the Magnetic Properties of Fe-Si-Cr SMCs

Haiming Long [1], Xiaojie Wu [2], Yunkun Lu [1], Haifeng Zhang [1] and Junjie Hao [1,*]

[1] Institute for Advanced Materials and Technology, University of Science and Technology Beijing, Beijing 100083, China; yingfengzui@163.com (H.L.); luyunkun123@126.com (Y.L.); 18503365756@163.com (H.Z.)

[2] Avic Chengdu Caic Electronics Co., Ltd., Chengdu 610091, China; wuxiaojie2020@163.com

* Correspondence: haojunjie@ustb.edu.cn

Abstract: Fe-Si-Cr soft magnetic powder cores (SMCs), with high electrical resistivity, magnetic permeability, saturation magnetic induction, and good corrosion resistance, are widely applied to inductors, filters, choke coils, etc. However, with the development of electronic technology with high frequency and high power density, the relative decline in the magnetic properties limits the high-frequency application of SMCs. In this paper, the phosphating process and polyimide (PI) insulation coating is applied to Fe-Si-Cr SMCs to reduce the core loss, including hysteresis loss and eddy current loss. The microstructure and composition of Fe-Si-Cr powders were analyzed by SEM, XRD, and Fourier-transform infrared spectra, respectively. The structural characteristics of the Fe-Si-Cr @ phosphate layer @ PI layer core–shell double coating were studied, and the best process parameters were determined through experiments. For SMCs with 0.4 wt% content of PI, the relative permeability is greater than 68%, and the core loss is the lowest, 7086 mW/cm^3; annealed at 500 °C, the relative permeability is greater than 57%, and the core loss is the lowest, 6222 mW/cm^3. A 0.4 wt% content of PI, annealed at 500 °C, exhibits the ideal magnetic properties: μ_e = 47 H/m, P = 6222 mW/cm^3.

Keywords: Fe-Si-Cr SMCs; PI; core–shell double coating; annealing; magnetic properties

Citation: Long, H.; Wu, X.; Lu, Y.; Zhang, H.; Hao, J. Effect of Polyimide-Phosphating Double Coating and Annealing on the Magnetic Properties of Fe-Si-Cr SMCs. *Materials* **2022**, *15*, 3350. https://doi.org/10.3390/ma15093350

Academic Editor: Alain Celzard

Received: 5 April 2022
Accepted: 29 April 2022
Published: 7 May 2022

Publisher's Note: MDPI stays neutral with regard to jurisdictional claims in published maps and institutional affiliations.

1. Introduction

Metal magnetic powder cores, belonging to a kind of soft magnetic composite material, are prepared by mixing ferromagnetic powder with an insulating medium [1]. They are commonly used in transformers, electronic communication, and radar due to the strengths of high saturation induction density, high magnetic permeability, and low total loss [2,3]. In recent years, Fe-Si-Cr SMCs, as new soft magnetic composite materials, have widely been applied to the inductance, filter, choking ring, and similar areas due to their higher electronic resistivity, permeability, and lower core loss in comparison with traditional silicon steel sheets, single-metal-based soft magnetic materials [4]. In high-frequency applications, adding Si can not only reduce the eddy current loss by increasing the resistivity of SMCs but also can form CrSi and CrSi$_2$ in Fe-Si-Cr SMCs with excellent temperature characteristics [5]. The addition of Cr can improve the mechanical strength, plasticity, and corrosion resistance of SMCs [6]. Compared with other iron-based SMCs, Fe-Si-Cr SMCs offer better broadband response characteristics and lower cost. Unfortunately, the increment in core loss as a result of increased operating frequency limits the large application of Fe-Si-Cr SMCs [7].

To reduce core loss P_{cv}, including hysteresis loss P_h and eddy current loss P_e, insulating coating and high-pressure forming are usually applied in the manufacturing process [8–18]. Generally, there are two types of coatings used to suppress eddy currents: organic coatings and inorganic coatings [19]. With the advantages of satisfactory adhesion and flexibility, organic substances such as epoxy resin [20] or phenolic resin [21] have been used as the insulating layer of SMCs. Due to high dislocation density and defects, high pressure causes an increase in hysteresis loss P_h. In order to eliminate defects such as lattice strain, a high-temperature annealing process is usually used. Some new characterization methods have promoted the study of SMCs [22–24]. However, the annealing process above 400 °C easily decomposes the organic resin [25]. Therefore, phosphate [26] and oxide [27,28] are used as the passivation layer of SMCs. However, the phosphate insulating layer will also collapse during the annealing process, resulting in a decrease in resistivity [26]. The organic coating has good adhesion but poor heat resistance. PI has higher heat resistance, insulation resistivity, and mechanical stability than ordinary resins, which is a potential organic coating material for magnetic powders. However, PI is a non-magnetic material; it can increase resistivity but decrease permeability [29]. To optimize the magnetic properties and reduce the core loss, high-temperature annealing is an effective method [30].

The process of preparing SMCs by powder metallurgy has been widely used to lower costs and improve efficiency [31,32]. This method is based on using each fine powder particle to make the insulating coating, which can significantly reduce the core loss of the SMCs. Therefore, the research on high-performance insulating coatings and coating methods for SMCs is currently a popular research subject [33]. Due to the excellent insulation performance of the organic coating but the decomposition temperature being low, it cannot be combined with subsequent high-temperature annealing treatment to eliminate the influence of residual stress during pressing. Therefore, the current research tends to use an inorganic coating to improve the annealing temperature [34]. There are few papers on the research of inorganic + organic double coating, especially research on using PI with high decomposition temperature as an organic coating. Although many inorganic materials also have excellent insulation properties and can significantly reduce eddy current loss, the compact density of inorganically coated SMCs is generally lower than that of those that are organically coated, so there will be defects, such as compact pores, which in turn affect magnetic properties, such as hysteresis loss.

In this paper, to reduce the core loss of SMCs at high frequencies, an inorganic phosphate + organic PI double coating with excellent insulation performance was used to improve powder resistivity and reduce eddy current loss; the organic coating PI with a higher decomposition temperature was used to increase the compact density and increase the annealing temperature to reduce the hysteresis loss. Finally, higher magnetic properties can be obtained by annealing in an argon atmosphere at 500 °C for 1 h.

2. Materials and Methods

2.1. Preparation of SMCs

Fe-Si-Cr powders with d50 = 10 μm were prepared by gas atomization, consisting of 3.3 wt% Si, 6.5 wt% Cr, and balance of Fe. The preparation process of Fe-Si-Cr @ phosphate layer @ PI layer core–shell double coating and SMCs is divided into three steps: phosphating, coating, and annealing, as shown in Figure 1. First, Fe-Si-Cr powders were pretreated with phosphate, as shown in Figure 1a. The phosphating procedure was carried out in 50 mL of acetone, mechanical stirring at room temperature for one hour followed by drying at 80 °C for one hour. Secondly, we prepared PI coating, as shown in Figure 1b. The Fe-Si-Cr powders were mixed uniformly with the various PI of 0 wt%, 0.4 wt%, 0.7 wt%, 1.0 wt%, respectively (the PI cannot be dissolved into the water or alcohol solution and can be dissolved with *N*-Methyl pyrrolidone), and air-dried at 80 °C for 12 h. The dry powders were passed through a screen of −100 mesh. Finally, we prepared and annealed the SMCs; as shown in Figure 1c, the coated Fe-Cr-Si powder was uniformly mixed with zinc stearate lubricant (0.6 wt%). Next, the coated powders were pressed into cores under applied axial stress of 600 MPa with outer diameter of 14 mm, inner diameter of 8 mm, and height of about 3 mm. Lastly, the SMCs were annealed at different temperatures from 300 °C to 500 °C for one hour in argon atmosphere with a pipe furnace.

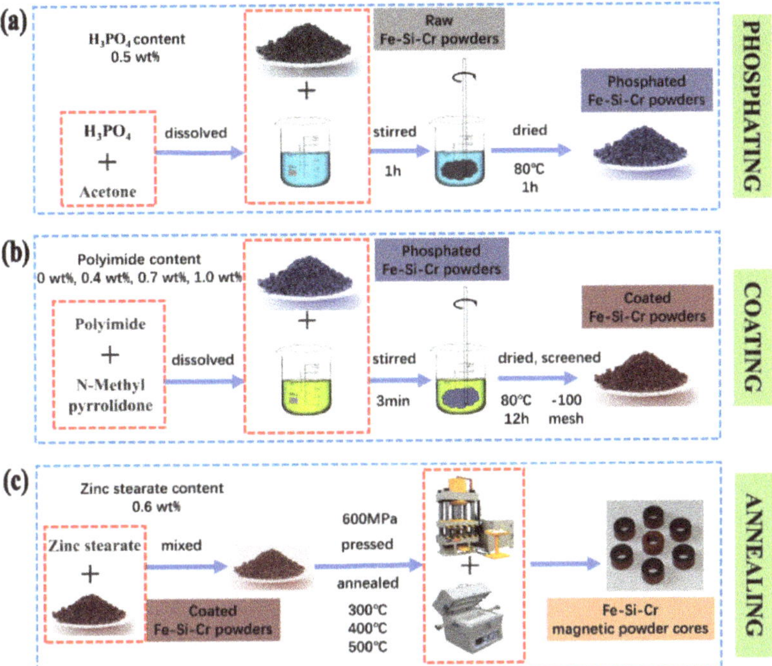

Figure 1. The preparation process of Fe-Si-Cr@PI core–shell structure coating and magnetic powder cores: (**a**) phosphating, (**b**) coating, and (**c**) annealing.

2.2. Test Method and Material Characterization

The inductance of the Fe-Si-Cr SMCs was measured by the LCR bridge tester, and we calculated the magnetic permeability by using Equation (1):

$$\mu_e = \frac{L \times 10^{-9} \times L_e}{4N A_e} \tag{1}$$

where μ_e is the effective permeability, L is the inductance of sample core, and L_e is the mean flux density path of the ring sample. N is the number of turns of the coil ($N = 25$), A_e is the area of cross-section. Figure 2 shows the magnetic powder core to be tested.

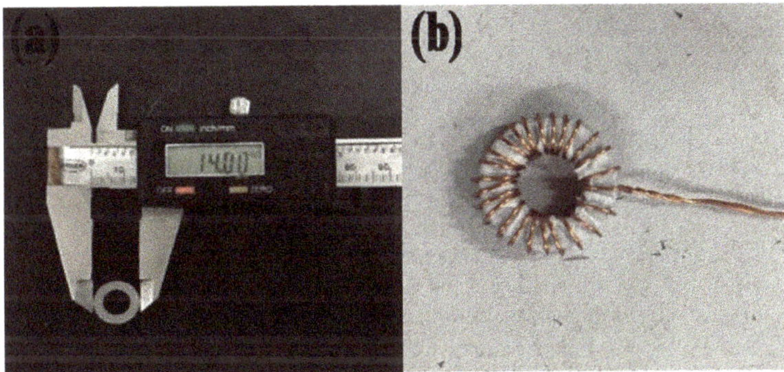

Figure 2. The SMCs sample to be tested: (**a**) magnetic powder core size and (**b**) coil winding before magnetic performance test.

The microstructure of uncoated and coated Fe-Si-Cr powder was characterized by scanning electron microscopy (SEM, LEO1450, CARL ZEISS, Oberkochen, Germany) equipped with the energy dispersive X-ray spectrometry (EDS, Quanta-200, CARL ZEISS, Oberkochen, Germany). FTIR was used to verify the phosphating effect and the coating effect of PI (Thermo Scientific Nicolet iS5, Thermo Fisher Scientific, Waltham, MA, USA). XRD was used to characterize the structure of the powder and SMCs (Rigaku Ultima IV, Rigaku Corporation, Tokyo, Japan). The kinetics of thermal decomposition of PI was investigated using synchronous thermal analyzer (TG-DSC, Q600, METTLER-TOLEDO, DE, USA). LCR bridge tester (TH2829C, Agitek, Xi'an, China) is used to measure the inductance of SMCs, the core loss was measured by an auto testing system for SMCs (IWATSU SY-943, IWATSU ELECTRIC, Tokyo, Japan) in the frequency range of 100 kHz^{-1} MHz, and the magnetic flux density was set to 50 mT.

3. Results

3.1. Characteristics of Phosphated and Coated Layer

After the two-step process of phosphating and coating, the oxide layer of the raw powder particles can be removed, and a certain thickness of the phosphate layer and PI insulation layer can be obtained, as shown in Figure 3. On the one hand, phosphating can remove the oxide layer on the surface of the original powder, including iron oxide, chromium oxide, and silicon oxide. On the other hand, a phosphating layer can be formed on the surface of powder in the phosphating process so as to increase the resistivity and reduce the eddy current loss [35]. PI is a non-magnetic material; it can increase resistivity but reduce permeability. To increase the insulation resistance without damaging the magnetic permeability, it is necessary to determine the appropriate content of PI addition—that is, to optimize the thickness of the Fe-Si-Cr @ phosphate layer @ PI layer core–shell double coating.

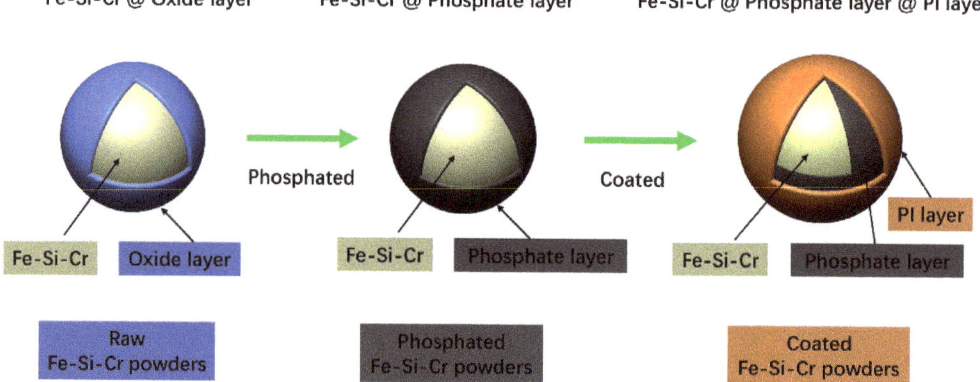

Figure 3. The effect of phosphating and coating process on the surface of Fe-Si-Cr powder particles.

3.1.1. Characteristics of the Phosphated Layer

The SEM images of the Fe-Si-Cr raw powder and the phosphated powder are shown in Figure 4. It can be seen from Figure 4a that the distribution of the Fe-Si-Cr raw powder particles is relatively dispersed, and most of the particles are spherical or nearly spherical (spindle shape). This is because the cooling rate of the gas atomization process is slower than that of the water atomization process, and it is easy to obtain a spherical powder. At the same time, for the surface oxide layer of powder particles, the gas-atomization process is much smaller than the water-atomization process. It can be seen from Figure 4b that the surface of the Fe-Si-Cr powder particles after phosphating is smooth, which indicates that the phosphate layer is evenly distributed on the surface of the powder. After the phosphating treatment, the phosphated substance—the reaction product of phosphoric acid, iron, and chromium—cannot be observed intuitively and is further characterized by other methods in a follow-up.

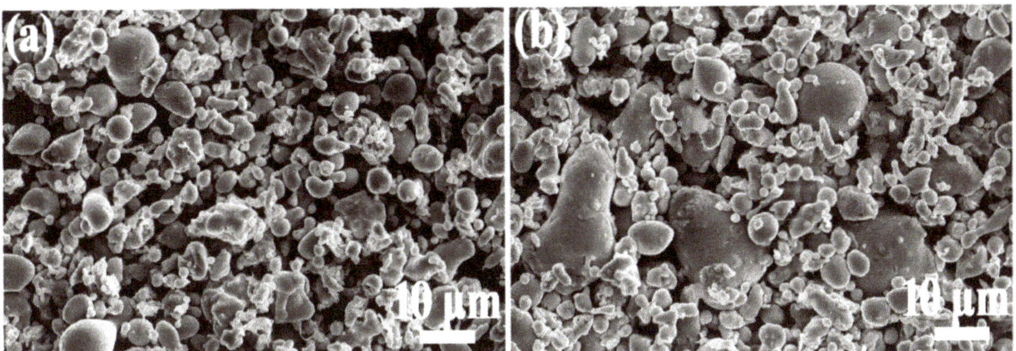

Figure 4. Comparison of Fe-Si-Cr powders: (**a**) Fe-Si-Cr raw powder, (**b**) Fe-Si-Cr phosphated powder.

The energy spectrum characteristics of the powder after phosphating are shown in Figure 5. It can be clearly observed that the *p* element is evenly distributed on the surface of the powder, which indicates that a phosphate layer is formed on the surface of the Fe-Si-Cr powder. The presence of phosphate can improve the resistivity of Fe-Si-Cr powder so as to ensure a relatively low eddy current loss and good processability [36].

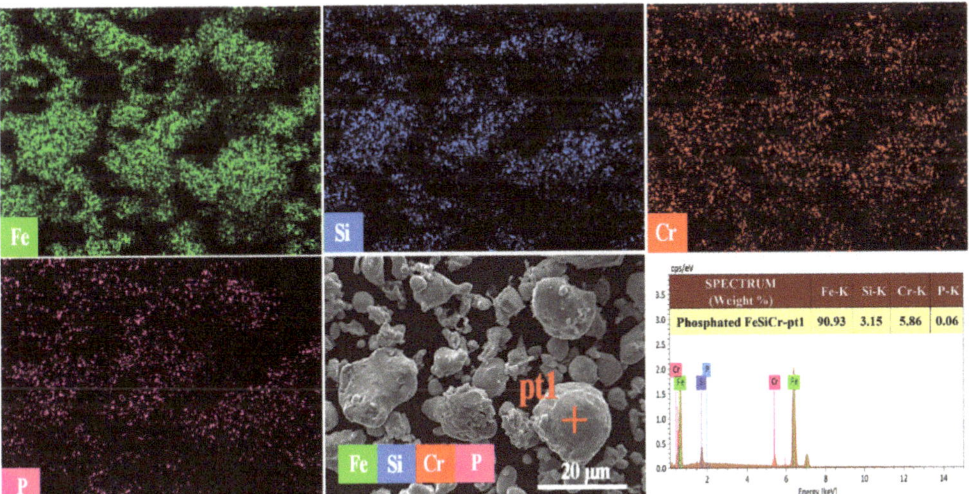

Figure 5. The EDS of Fe-Si-Cr phosphated powder.

3.1.2. Characteristics of the PI Coating Layer

Figure 6 is a microscopic image of Fe-Si-Cr powder coated with different content of PI. In Figure 6a, the Fe-Si-Cr powder is uncoated after phosphating. The powder is relatively dispersed and has an average particle size of 10 μm. In Figure 6b,c, the Fe-Si-Cr powder coated with PI is a mostly irregular, spherical powder. Meanwhile, there is a reunion as a result of the bonding effect of PI.

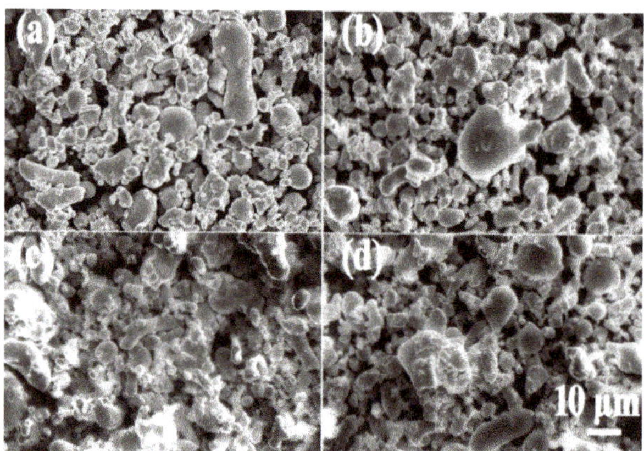

Figure 6. Fe-Si-Cr powders coated with PI: (**a**) 0 wt% PI, (**b**) 0.4 wt% PI, (**c**) 0.7 wt% PI, (**d**) 1.0 wt% PI.

Figure 7 shows the Fourier-transform infrared spectrum. It can be seen from the Fe-Si-Cr raw powder that the broad absorption peak at 3438 cm^{-1} is the -OH stretching vibration of adsorbed water, and the absorption peaks at 2928 cm^{-1} and 2855 cm^{-1} are the symmetric and asymmetric stretching vibrations of -CH in the methylene group. The absorption peak at 1626 cm^{-1} is the -OH bending vibration of water molecules, the absorption peak at 1110 cm^{-1} is the asymmetric stretching vibration of Si-O-Si or Fe-O-Si, and the absorption peak at 663 cm^{-1} is caused by the stretching vibration of Cr-O. For the phosphated powder, new absorption peaks appear at 567 cm^{-1} and 802 cm^{-1}; the absorption peak at 567 cm^{-1} is the bending vibration of O-P-O and the asymmetric stretching vibration of P-O at 802 cm^{-1}. According to these two absorptions, the presence of the peak can determine that the sample contains PO_4^{-3} and the intensity of the absorption peak at 1112 cm^{-1} becomes lower. It is possible that phosphoric acid interacts with Fe-O-Si, which reduces its content. For the phosphated and coated powder, new absorption peaks appeared at 1725 cm^{-1}, 1387 cm^{-1}, 1250 cm^{-1}, and 724 cm^{-1}. These absorption peaks are all caused by the characteristic peaks of PI. Among them, 1725 cm^{-1} is carbonyl C=O stretching vibration, 1387 cm^{-1} is C-N stretching vibration, 1250 cm^{-1} is C-O stretching vibration, and 720 cm^{-1} is C-O bending vibration, indicating that the powder is successfully coated with PI [37–39].

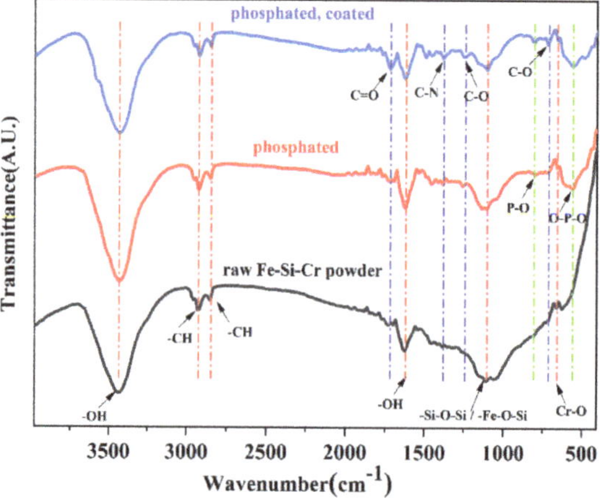

Figure 7. FTIR spectrum of uncoated and coated Fe-Si-Cr powder (from bottom to top is the raw powder, phosphated powder, and phosphate-coated powder.

3.2. Effect of PI on the Phase Composition and Magnetic Properties

3.2.1. Effect of PI Content on the Phase Composition of the Fe-Si-Cr SMCs

Figure 8 is the XRD pattern of Fe-Si-Cr powders with different PI coating amounts. Three sharp characteristic peaks (110), (200), and (211) are detected, which are consistent with the peaks of the α-Fe and Fe$_3$Si [40,41]. The Si and Cr atoms are solid-dissolved in the crystal lattice of α-Fe, and a solid solution of bcc -α-Fe (Si, Cr) is formed. It can be seen from the figure that the characteristic peak intensity after phosphating is significantly lower than the characteristic peak intensity before phosphating. This is due to the reaction of iron and phosphoric acid to form a phosphate layer, which reduces the characteristic peak intensity of α-Fe [42], and FTIR spectroscopy analysis also confirmed the existence of the phosphate layer. As the amount of PI coating increases from 0 to 1.0 wt%, the characteristic peak intensity also shows a downward trend. This is because the thickening of the PI layer weakens the X-ray absorption of the Fe-Si-Cr matrix. However, due to the thinner phosphate layer and PI layer, the XRD failed to detect the phosphide and PI phases.

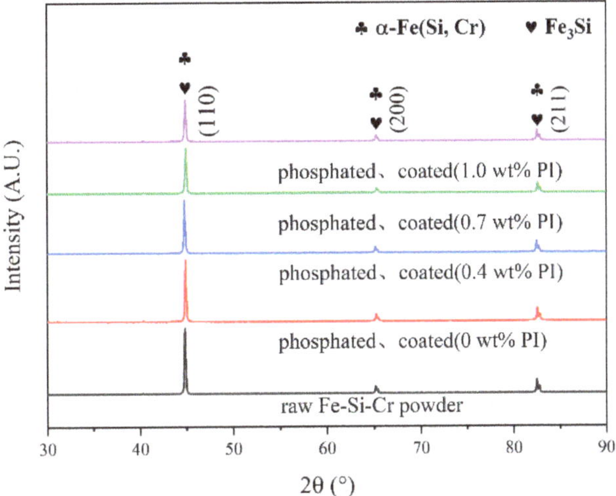

Figure 8. XRD patterns of Fe-Si-Cr powders with different PI coating contents.

3.2.2. The Trend of Magnetic Properties with PI Content

The magnetic properties of SMCs can be characterized by relative permeability and DC bias capability. DC bias refers to the superposition of an alternating current when an alternating magnetic field and DC magnetic field are simultaneously applied to the magnetic core. Figure 9a shows the DC bias capacity curve of Fe-Si-Cr SMCs coated with different contents of PI. It can be seen from the figure that the DC bias capacity of SMCs without PI is the best. When the applied magnetic field strength is 100 Oe, the relative permeability reaches 75%. Compared with the sample without PI, PI reduces the relative permeability of SMCs; however, when the magnetic field strength is 100 Oe and the PI content is 0.4 wt%, the relative permeability is >68%, indicating that its DC bias ability is not poor. In the range of 0~1.0 wt%, the relative permeability increases with the increase in PI content. This is because, in the applied DC magnetic field, SMCs are magnetized, the pressing density of SMCs without PI is low, and the air gap hinders the rotation and displacement of the magnetic domain, which makes it difficult for SMCs to be magnetized to saturation. However, the compaction density of SMCs with PI is relatively high, the air gap is small, and the displacement and rotation of the magnetic domain are relatively small, so it is easier to be magnetized to saturation. However, the addition of non-magnetic PI resin reduces the proportion of magnetic substances in SMCs. At this time, the resin hinders the rotation and displacement of magnetic domains, such as the gap between particles. Therefore, the relative permeability decreases compared with that without PI resin.

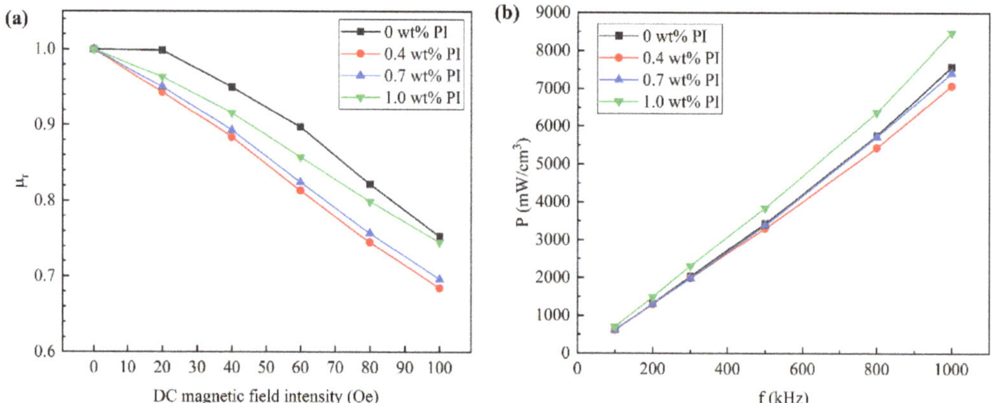

Figure 9. The trend of magnetic properties with PI content: (**a**) relative permeability; (**b**) core loss.

Figure 9b is the core loss curve of Fe-Si-Cr SMCs coated with different PI contents. The total loss (P_{cv}) is composed of hysteresis loss (P_h), eddy current loss (P_e), and residual loss (P_c). The residual loss is the micro-eddy current generated by the magnetic domain wall, which is very small compared to the hysteresis loss and eddy current loss. It can be ignored. The core loss of all the samples increases with the increase in frequency; at the same frequency, the SMCs have the smallest core loss at the 0.4 wt% PI. When the frequency is 1000 Hz, the core loss is 7086 mW/cm^3. Due to the application of double-insulating coatings, the coated cores exhibit lower magnetic loss than uncoated cores; insulating coated layers effectively hinder the current of intra-particles and inter-particles and thus reduce the eddy current loss. When the PI content is 1.0 wt%, the core loss is the largest. Because the insulating layer is a non-magnetic substance, it acts as a hindrance in the magnetization process of the Fe-Si-Cr SMCs, resulting in an increase in hysteresis loss. When the insulating layer is thicker, the eddy current loss is reduced, while the hysteresis loss is increased so that the total core loss is increased.

3.3. The Effect of Annealing on the Phase Composition and Magnetic Properties

3.3.1. The Choice of the Annealing Temperature and the Change of the Phase Composition

From the point of view of eliminating residual stress, the higher the annealing temperature, the better the effect. However, the selection of the annealing temperature should consider the influence of the passivation layer and coated layer. Phosphating treatment will form a passivation layer on the surface of the powder, and the phosphating layer will crystallize with iron at 500 °C [43]. In addition, the final annealing temperature should be determined in combination with the heat resistance of the PI coating. As shown in Figure 10, according to the DSC curve, PI has an endothermic peak near 607 °C, which is the thermal decomposition temperature of PI. According to the TG curve, the 2% thermal weight loss temperature is as high as 500 °C (mainly due to the evaporation of adsorbed water in PI powder), and the maximum heat-resistant temperature of PI can reach 600 °C. Therefore, the final annealing temperature range is determined as 300~500 °C.

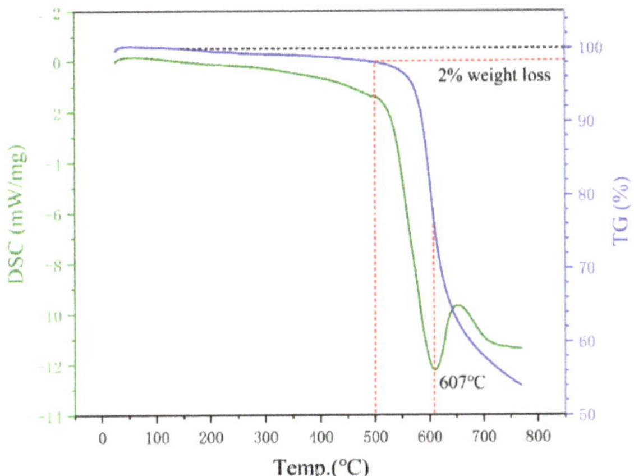

Figure 10. Kinetic curves of thermal decomposition of PI in argon.

Figure 11 is the XRD pattern of Fe-Si-Cr SMCs with different annealing temperatures. Similar to Fe-Si-Cr powder, three sharp characteristic peaks (110), (200), and (211) are detected, and the phase composition is mainly α-Fe (Si, Cr) solid solution and Fe_3Si. It can be seen that annealing only eliminates the internal stress of SMCs without changing their phase composition.

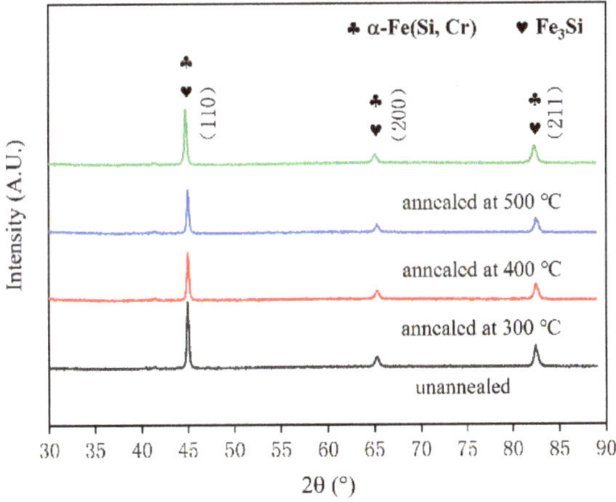

Figure 11. XRD patterns of Fe-Si-Cr SMCs at different annealing temperatures.

3.3.2. The Trend of Magnetic Properties with Annealing Temperature

The pressing process will reduce the air gap and produce residual stress in the SMCs before the annealing process. Due to the reduction in non-magnetic materials and the increase in effective permeability, the DC bias ability becomes worse. The trend of DC bias capacity with heat treatment is shown in Figure 12a; the relative permeability of the SMCs annealed at 300 °C is the highest, reaching 69% at 100 Oe. The relative permeability of cores reduced gradually with the decrease in the annealing temperature from 300 °C to 500 °C. This is because the higher the annealing temperature, the lower the domain wall resistance, and the corresponding magnetic core is easily magnetized to saturation. However, under a 100 Oe magnetic field intensity, the magnetic permeability of the SMCs annealed at 500 °C reaches 57%, which also does not show a poor DC bias.

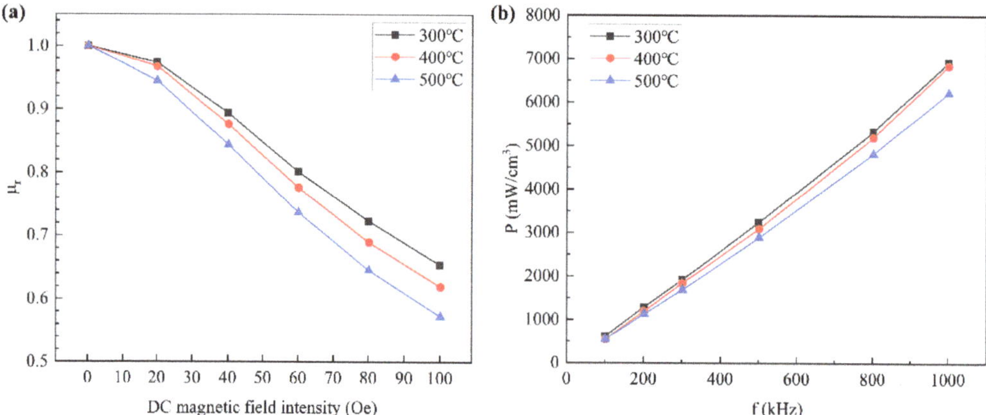

Figure 12. The trend of magnetic properties with annealing temperature (the powder was phosphated and coated with 0.4 wt% PI): (**a**) relative permeability; (**b**) core loss.

Annealing can eliminate the residual internal stress and dislocation generated after magnetic particle pressing, compact the structure, reduce air and other defects, reduce the hysteresis loss coefficient, and finally, reduce the hysteresis loss. The higher the annealing temperature, the more thorough the removal of internal stress, air and dislocation, and other defects between magnetic particles, and the more obvious the effect of loss reduction [44,45]. Figure 12b shows the core loss of Fe-Si-Cr SMCs annealed at different temperatures. It can be seen from the figure that the iron loss decreases gradually with the increase in annealing temperature. The core annealed at 500 °C has the lowest loss, which is only 6222 mW/cm^3 at 1000 Hz.

3.4. The Effect of PI Content and Annealing Temperature on Effective Permeability

Figure 13 shows the effective permeability of Fe-Si-Cr SMCs at different PI contents and different annealing temperatures. It can be seen that all the samples show the same trend; with the increase in PI content, the effective permeability increases first and then decreases at the same temperature. The phenomenon can be ascribed that the increase in PI leads to the high compaction density of the powder, and thereby the air gap decreases and the effective permeability increases, which can be confirmed in Table 1. The density of core coated at 0.4 wt% PI and annealed at 500 °C reached 6.213 g/cm^3. However, PI is a non-magnetic substance, and the increase in PI causes a decrease in magnetic material and a decrease in effective permeability. The best response was obtained for the sample coated with 0.4 wt% PI. During annealing, the atomic disorder state was changed to an ordered state, the microstructure of Fe-Si-Cr SMCs was optimized well, the air gap was reduced, and the annealed cores were denser, and it can be seen in Table 1 that the density of cores increased with the increase in annealing temperature. Thus, improving the effective permeability of Fe-Si-Cr SMCs, the SMCs show an ideal effect at an annealing temperature of 500 °C. The Fe-Si-Cr SMCs, with 0.4 wt% content of PI and heat treatment temperature at 500 °C, exhibited the best magnetic properties: $\mu_e = 47$ H/m, $p = 6222$ mW/cm^3.

Table 1. Density tests of Fe-Si-Cr SMCs coated with various content of PI at different annealing temperatures.

ρ (g/cm³)		FeCrSi/PI Content (wt%) in This Study				FeCrSi/Sodium Silicate [40]
		0	0.4	0.7	1.0	
Temperature (°C)	300	5.917	6.130	6.066	5.869	/
	400	6.051	6.206	6.146	5.920	/
	450	/	/	/	/	5.610
	500	6.163	6.213	6.179	6.112	/
	550	/	/	/	/	6.020
	650	/	/	/	/	6.140

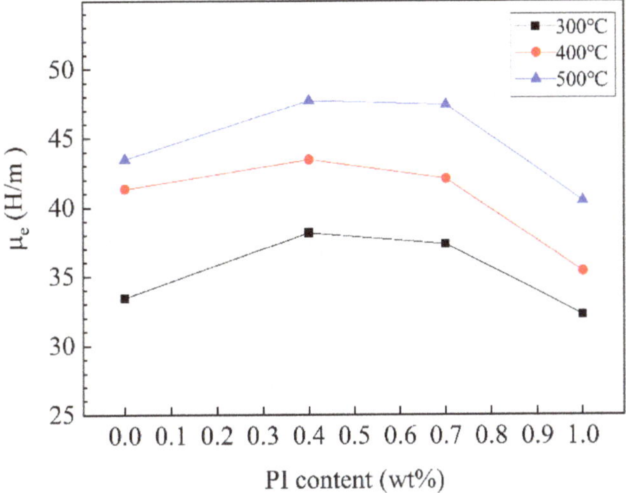

Figure 13. Trend of PI content and annealing temperature on effective permeability of Fe-Si-Cr SMCs.

In the field of electromagnetism, the total core loss (P_{cv}) consists of hysteresis loss (P_h), eddy current loss (P_e), and residual loss (P_c); the residual loss is the micro-eddy current generated by the domain wall, which is very small compared with the hysteresis loss and eddy current loss and can be ignored. Additionally, the total core loss P_{cv} can be expressed as Equation (2) [14,17].

$$P_{cv} = P_h + P_e = K_h \times f + k_e \times f^2 \tag{2}$$

where K_h is the hysteresis loss coefficient, K_e is the eddy current loss coefficient, and f is the frequency. At low frequencies, the increase in total loss is mainly the increase in hysteresis loss, while at medium and high frequencies, the increase in total loss is mainly eddy current loss. The comparison of magnetic properties between this study and the literature is shown in Table 2. In this study, the effects of inorganic + organic double coating and heat treatment on the total core loss of SMCs are preliminarily explored; however, more accurate quantitative research on hysteresis loss and eddy current loss has not been completed. In further research, the quantitative results of the influence on each component of core loss P_{cv} will be emphatically considered, and the effects of different process steps, including powder coating preparation, pressing, and annealing on hysteresis loss P_h and eddy current loss P_e will be evaluated so as to provide guidance for industrial production. In addition, the use of organic PI coating can significantly improve the corrosion resistance of SMCs, which is also worthy of further research.

Table 2. Comparison of the magnetic performances in this study and the literature.

Samples	μ_e (H/m)	P_{cv} (mW/cm^3)			
		0.02T 1000 kHz	0.05T 100 kHz	0.05T 500 kHz	0.05T 1000 kHz
In this study	47.5	/	547	2888	6222
FeSiCr/phosphate [36]	44.5	/	780	/	/
FeSiCr/PA6 [42]	18	/	1100	/	/
FeSiCr/MnZnFe [18]	55	/	738	/	/
FeSiCr/yttrium nitrate [32]	41	600	/	/	6250
FeSiCr/carbonyl iron [34]	37	560	/	/	/
FeCrSi/sodium silicate [40]	34.9	/	/	/	/

4. Conclusions

In this paper, to reduce the core loss of SMCs in high-frequency application environments, the strategies of inorganic–organic double-insulating coating and high-temperature annealing were adopted. Phosphating can not only remove the oxide layer on the powder surface but also form a phosphate insulating coating on the powder surface. This insulating coating can significantly reduce the core loss of SMCs. At the same time, this phosphate is also a good intermediate transition layer for coating organic PI; SMCs coated with organic PI can significantly reduce the core loss, the addition of PI can increase the lubricity of powder in the pressing stage of SMCs, cause the pressed compact have high density, and reduce the existence of defects such as pores, which is conducive to reducing the core loss. Annealing is an effective method to reduce the influence of the pressing process on the magnetic properties of SMCs, which can significantly reduce the core loss. In this study, the Fe-Si-Cr SMCs, with 0.4 wt% content of PI and annealing temperature at 500 °C, exhibit the best magnetic properties: μ_e = 47 H/m, p = 6222 mW/cm^3.

Author Contributions: Conceptualization, H.L. and J.H.; methodology, H.L. and J.H.; validation, H.L. and X.W.; formal analysis, H.L. and H.Z.; investigation, H.L., X.W. and H.Z.; resources, J.H. and Y.L.; data curation, H.L. and X.W.; writing—original draft preparation, H.L.; writing—review and editing, H.L.; visualization, Y.L.; supervision, J.H.; project administration, J.H. All authors have read and agreed to the published version of the manuscript.

Funding: This research received no external funding.

Institutional Review Board Statement: Not applicable.

Informed Consent Statement: Not applicable.

Data Availability Statement: Not applicable.

Conflicts of Interest: The authors declare no conflict of interest.

References

1. Xu, L.; Yan, B. Fe–6.5% Si/SiO$_2$ powder cores prepared by spark plasma sintering: Magnetic properties and sintering mechanism. *Int. J. Mod. Phys. B* **2017**, *31*, 1744011. [CrossRef]
2. Hosseinzadeh, S.; Elahi, P.; Behboudnia, M.; Sheikhi, M.H.; Mohseni, S.M. Structural and magnetic study of metallo-organic YIG powder using 2-ethylhexanoate carboxylate-based precursors. *Mod. Phys. Lett. B* **2019**, *33*, 1950100. [CrossRef]
3. Xia, C.; Peng, Y.; Yi, Y.; Deng, H.; Zhu, Y.; Hu, G. The magnetic properties and microstructure of phosphated amorphous FeSiCr/silane soft magnetic composite. *J. Magn. Magn. Mater.* **2019**, *474*, 424–433. [CrossRef]
4. Li, Z.B.; Liu, J. Preparation of Fe Si Cr alloy soft magnetic powder by gas atomization and its morphology analysis. *World Nonf. Metals.* **2017**, *14*, 9–10.
5. Xu, H.P.; Wang, R.W.; Wei, D.; Zeng, C. Crystallization kinetics and magnetic properties of FeSiCr amorphous alloy powder cores. *J. Magn. Magn. Mater.* **2015**, *385*, 326–330. [CrossRef]
6. Zou, B.; Zhou, T.; Xin, W.; Song, T. Influence of Cr Content on Electronic Structures and Electromagnetic Properties of FeSiCr Powders. *Rare. Metal. Mat. Eng.* **2013**, *42*, 313–316.
7. Ren, J.; Bo, L.I.; Wang, J.; Pang, X.; Guo, H. Preparation and magnetic properties of FeSiCr/SiO$_2$ soft magnetic composites. *Electron. Comp. Mater.* **2018**, *37*, 51–56. [CrossRef]

8.	Hsiang, H.I.; Wang, S.K.; Chen, C.C. Electromagnetic properties of FeSiCr alloy powders modified with amorphous SiO_2. *J. Magn. Magn. Mater.* **2020**, *514*, 167151. [CrossRef]

9.	Guo, R.; Wang, S.; Yu, Z.; Sun, K.; Lan, Z. FeSiCr@NiZn SMCs with ultra-low core losses, high resistivity for high frequency applications. *J. Alloys Compd.* **2020**, *830*, 154736. [CrossRef]

10.	Nie, W.; Yu, T.; Wang, Z.; Wei, X. High-performance core-shell-type FeSiCr@MnZn soft magnetic composites for high-frequency applications. *J. Alloys Compd.* **2020**, *864*, 158215. [CrossRef]

11.	Li, Z.; Dong, Y.; Pauly, S.; Chang, C.; Wei, R.; Li, F.; Wang, X.-M. Enhanced soft magnetic properties of Fe-based amorphous powder cores by longitude magnetic field annealing. *J. Alloys Compd.* **2017**, *706*, 1–6. [CrossRef]

12.	Geng, K.; Xie, Y.; Xu, L.; Yan, B. Structure and magnetic properties of ZrO_2 -coated Fe powders and Fe/ZrO_2 soft magnetic composites. *Adv. Powder. Technol.* **2017**, *28*, 2015–2022. [CrossRef]

13.	Wang, L.; Xiong, H.; Rehman, S.U.; Chen, Y.; Tan, Q.; Zhang, L.; Zhong, M.; Zhong, Z. Optimized microstructure and impedance matching for improving the absorbing properties of core-shell $C@Fe_3C/Fe$ nanocomposites. *J. Alloys Compd.* **2019**, *780*, 552–557. [CrossRef]

14.	Yaghtin, M.; Taghvaei, A.H.; Hashemi, B.; Janghorban, K. Effect of heat treatment on magnetic properties of iron-based soft magnetic composites with Al_2O_3 insulation coating produced by sol-gel method. *J. Alloys Compd.* **2013**, *581*, 293–297. [CrossRef]

15.	Peng, Y.; Nie, J.; Zhang, W.; Jian, M.; Bao, C.; Yang, C. Effect of the addition of Al_2O_3 nanoparticles on the magnetic properties of Fe soft magnetic composites. *J. Magn. Magn. Mater.* **2016**, *399*, 88–93. [CrossRef]

16.	Wang, J.; Fan, X.; Wu, Z.; Li, G. Intergranular insulated Fe/SiO_2 soft magnetic composite for decreased core loss. *Adv. Powder. Technol.* **2016**, *27*, 1189–1194. [CrossRef]

17.	Yang, B.; Li, X.; Guo, R.; Yu, R. Oxidation fabrication and enhanced soft magnetic properties for core-shell $FeCo/CoFe_2O_4$ micron-nano composites. *Mater. Des.* **2017**, *121*, 272–279. [CrossRef]

18.	Mori, S.; Mitsuoka, T.; Sugimura, K.; Hirayama, R.; Sonehara, M.; Sato, T.; Matsushita, N. Core-shell structured Mn-Zn-Fe ferrite/Fe-Si-Cr particles for magnetic composite cores with low loss. *Adv. Powder. Technol.* **2018**, *29*, 1481–1486. [CrossRef]

19.	Yaghtin, M.; Taghvaei, A.H.; Hashemi, B.; Janghorban, K. Structural and magnetic properties of $Fe-Al_2O_3$ soft magnetic composites prepared by sol-gel method. *Int. J. Mater. Res.* **2013**, *105*, 474–479. [CrossRef]

20.	Xiao, L.; Sun, Y.; Ding, C.; Yang, L.; Yu, L. Anneling effects on magnetic properties and strength of organic-silicon epoxy resin-coated soft magnetic composites. *J. Mech. Eng. Sci.* **2014**, *228*, 2049–2058. [CrossRef]

21.	Kollár, P.; Birčáková, Z.; Füzer, J.; Bureš, R.; Faberova, M. Power loss separation in Fe-based composite materials. *J. Magn. Magn. Mater.* **2013**, *327*, 146–150. [CrossRef]

22.	Baco-Carles, V.; Huguet, T.; Llibre, J.; Baylac, V.; Pasquet, I.; Tailhades, P. Laser powder bed fusion applied to the manufacture of bulk or structured magnetic cores. *J. Mater. Res. Technol.* **2022**, *18*, 599–610. [CrossRef]

23.	Ammarullah, M.I.; Afif, I.Y.; Maula, M.I.; Winarni, T.I.; Tauviqirrahman, M.; Akbar, I.; Basri, H.; van der Heide, E.; Jamari, J. Tresca Stress Simulation of Metal-on-Metal Total Hip Arthroplasty during Normal Walking Activity. *Materials* **2021**, *14*, 7554. [CrossRef] [PubMed]

24.	Jamari, J.; Ammarullah, M.I.; Saad, A.P.M.; Syahrom, A.; Uddin, M.; van der Heide, E.; Basri, H. The Effect of Bottom Profile Dimples on the Femoral Head on Wear in Metal-on-Metal Total Hip Arthroplasty. *J. Funct. Biomater.* **2021**, *12*, 38. [CrossRef]

25.	Shokrollahi, H.; Janghorban, K. Soft magnetic composite materials (SMCs). *J. Mater. Process. Technol.* **2012**, *189*, 1–12. [CrossRef]

26.	Hsiang, H.I.; Fan, L.F.; Hung, J.J. Phosphoric acid addition effect on the microstructure and magnetic properties of iron-based soft magnetic composites. *J. Magn. Magn. Mater.* **2018**, *447*, 1–8. [CrossRef]

27.	Jiang, H.Y.; Zhong, W.; Wu, X.L.; Tang, N.J.; Liu, W.; Du, Y.W. Direct and alternating current magnetic properties of FeNi particles coated with SiO_2. *J. Alloys Compd.* **2004**, *384*, 264–267. [CrossRef]

28.	Fan, X.; Wang, J.; Wu, Z.Y.; Li, G. Core–shell structured $FeSiAl/SiO_2$ particles and Fe_3Si/Al_2O_3 soft magnetic composite cores with tunable insulating layer thicknesses. *Mat. Sci. Eng. B-Adv.* **2015**, *201*, 79–86. [CrossRef]

29.	Qian, K.M.; Ji, S.; Wu, M.; Zhang, Y.S.; Gong, C.H. Effect of insulation material on the properties of nanocrystal soft magnetic alloy magnetic powder Core. *Metal. Funct. Mater.* **2004**, *5*, 10–12. [CrossRef]

30.	Ye, C.; Huang, J.; Li, Q. Effect of heat treatment on density and magnetic properties of warm compacted FeSiAl magnetic powder cores. *Heat Treat. Met.* **2017**, *42*, 151–154. [CrossRef]

31.	Yu, H.; Zhou, S.; Zhang, G.; Dong, B.; Meng, L.; Li, Z.; Dong, Y.; Cao, X. The phosphating effect on the properties of FeSiCr alloy powder. *J. Magn. Magn. Mater.* **2022**, *552*, 168741. [CrossRef]

32.	Hsiang, H.; Fan, L.; Ho, K. Minor yttrium nitrate addition effect on FeSiCr alloy powder core electromagnetic properties. *J. Magn. Magn. Mater.* **2017**, *444*, 1–6. [CrossRef]

33.	Zhou, B.; Dong, Y.; Liu, L.; Chi, Q.; Zhang, Y.; Chang, L.; Bi, F.; Wang, X. The core-shell structured Fe-based amorphous magnetic powder cores with excellent magnetic properties. *Adv. Powder Technol.* **2019**, *30*, 1504–1512. [CrossRef]

34.	Xia, C.; Peng, Y.; Yi, X.; Yao, Z.; Zhu, Y.; Hu, G. Improved magnetic properties of FeSiCr amorphous soft magnetic composites by adding carbonyl iron powder. *J. Non-Cryst. Solids* **2021**, *559*, 120673. [CrossRef]

35.	Cai, P.P.; Liang, J.K.; Xu, W.Z.; Wang, Z.N.; Gong, Y.H.; Shen, B.B. Effect of Phosphating Process on the Performance of FeSiCr Soft Magnetic Powders. *Mech. Res. Appl.* **2017**, *5*, 58–64. [CrossRef]

36.	Wu, X.J.; Chen, C.G.; Hao, J.J.; Zhao, T.C.; Ren, Z.K. Effect of Phosphating and Heat Treatment on Magnetic Properties of Fe-3.3Si-6.5Cr Soft Magnetic Composites. *J. Supercond. Nov. Magn.* **2020**, *33*, 1889–1897. [CrossRef]

37. Morimune-Moriya, S.; Obara, K.; Fuseya, M.; Katanosaka, M. Development and characterization of strong, heat-resistant and thermally conductive polyimide/nanodiamond nanocomposites. *Polymer* **2021**, *230*, 124098. [CrossRef]
38. Ke, H.; Zhao, L.; Zhang, X.; Qiao, Y.; Wang, X. Performance of high-temperature thermosetting polyimide composites modified with thermoplastic polyimide. *Polym. Test.* **2020**, *90*, 106746. [CrossRef]
39. Zhao, X.; Wu, Q.; Zhang, S.; Wei, H.; Wang, R.; Wang, C. Synthesis, processability and photoluminescence of pyrene-containing polyimides. *J. Mater. Res. Technol.* **2020**, *9*, 14599–14608. [CrossRef]
40. Chen, S.F.; Chang, H.Y.; Wang, S.J.; Chen, S.H.; Chen, C.C. Enhanced electromagnetic properties of Fe–Cr–Si alloy powders by sodium silicate treatment. *J. Alloys Compd.* **2015**, *637*, 30–35. [CrossRef]
41. Yi, X.W.; Li, Q.B.; Peng, Y.D.; Zhao, Y.F.; Zhu, S.Z. Effect of Processing Condition on Microstructure and Properties of FeSiAl Powder Coated with Metal Oxides by Using a NaOH Solution. *J. Supercond. Nov. Magn.* **2021**, *34*, 2957–2968. [CrossRef]
42. Wang, L.; Qiao, L.; Zheng, J.; Cai, W.; Ying, Y.; Li, W.; Che, S.; Yu, J. Microstructure and properties of FeSiCr/PA6 composites by injection molding using FeSiCr powders by phosphating and coupling treatment. *J. Magn. Magn. Mater.* **2017**, *452*, 210–218. [CrossRef]
43. Hu, C.Z.; Andrade, J.D. Pyrolyzed, conducting kapton polyimide: An electrically conducting material. *J. Appl. Polym. Sci.* **1985**, *30*, 4409–4415. [CrossRef]
44. Boglietti, A.; Cavagnino, A.; Ferraris, L.; Lazzari, M. The annealing influence onto the magnetic and energetic properties in soft magnetic material after punching process. *IEEE. Int. Electr. Mach. Drive. Conf.* **2003**, *1*, 503–508. [CrossRef]
45. Chicinas, I.; Geoffroy, O.; Isnard, O.; Pop, V. Soft magnetic composite based on mechanically alloyed nanocrystalline Ni$_3$Fe phase. *J. Magn. Magn. Mater.* **2005**, *290*, 1531–1534. [CrossRef]

materials

MDPI

Article

A Study on the Static Magnetic and Electromagnetic Properties of Silica-Coated Carbonyl Iron Powder after Heat Treatment for Improving Thermal Stability

Xu Yan, Xinyuan Mu, Qinsheng Zhang, Zhanwei Ma, Chengli Song * and Bin Hu *

State Key Laboratory for Oxo Synthesis and Selective Oxidation, Lanzhou Institute of Chemical Physics, Chinese Academy of Sciences, Lanzhou 730000, China; yanxu@licp.cas.cn (X.Y.); muxinyuan@licp.cas.cn (X.M.); zhangqinsheng@licp.cas.cn (Q.Z.); mazhanwei@licp.cas.cn (Z.M.)
* Correspondence: songchl@licp.cas.cn (C.S.); hcom@licp.cas.cn (B.H.)

Abstract: In order to study the thermal stability of coated carbonyl iron powder (CIP) and its influence on magnetic properties, carbonyl iron powder was coated with a silica layer and then annealed in an air atmosphere at elevated temperatures. Transmission electron microscopy (TEM) analysis and Fourier transform infrared spectroscopy confirmed the existence of a silicon dioxide layer with a thickness of approximately 80~100 nm. Compared with uncoated CIP, the silicon-coated CIP still maintained a higher absorption performance after annealing, and the calculated impedance matching value Z only slightly decreased. It is worth noting that when the annealing temperature reached 300 °C, coercivity (H_c) increased, and the real and imaginary parts of the permeability decreased, which means that the silicon dioxide layer began to lose its effectiveness. On the contrary, the significant decrease in microwave absorption ability and impedance matching value Z of uncoated CIP after annealing were mainly because the newly formed oxide on the interface became the active polarization center, leading to an abnormal increase in permittivity. In terms of the incremental mass ratio after annealing, 2% was a tipping point for permeability reduction.

Keywords: microwave absorption; thermal stability; silica coating; carbonyl iron powders

Citation: Yan, X.; Mu, X.; Zhang, Q.; Ma, Z.; Song, C.; Hu, B. A Study on the Static Magnetic and Electromagnetic Properties of Silica-Coated Carbonyl Iron Powder after Heat Treatment for Improving Thermal Stability. *Materials* **2022**, *15*, 2499. https://doi.org/10.3390/ma15072499

Academic Editor: Anton Nikiforov

Received: 12 February 2022
Accepted: 21 March 2022
Published: 28 March 2022

Publisher's Note: MDPI stays neutral with regard to jurisdictional claims in published maps and institutional affiliations.

1. Introduction

In recent years, with the rapid development of wireless communication technology and high-frequency devices, microwave-absorbing materials have attracted more and more attention in the military and civilian fields [1,2]. In actual use, designing the wave absorber into a different shape or a coating filled with magnetic metal particles plays a vital role in wave absorption. The most commonly used magnetic metal particles are carbonyl iron particles (CIPs), characterized by high saturation magnetization (M_s), uniform spheres, smaller than 10 microns, a narrow particle size distribution, good microwave absorption, and higher cost-effectiveness applications. However, carbonyl iron is a highly reactive chemical substance, and it is easily oxidized by oxygen in the presence of water [3] or a temperature higher than 200 °C [4]. As a result, as the carbonyl iron particles are gradually oxidized, the electromagnetic performance drastically deteriorates. This shortcoming limits the post-treatment of carbonyl iron powder under high-temperature or complex environments.

To solve this problem, covering materials are used to coat the iron particles, which insulate the interaction of oxygen and iron cores, thereby preventing the oxidation of the carbonyl iron powder. In the past decades, various heat-resisting materials have been used to coat iron powders for improvement of thermal stability, such as aluminum phosphate [5], polyaniline [6], Al_2O_3 [7], Al [8,9], Ag [10], Co [11], Ni [12], and silica [13–16]. These materials improved the thermal stability of the coated samples to varying degrees. However, few works were devoted to studying the variation of the samples' static magnetic

---

and electromagnetic properties before and after heat treatment, which is very important for practical applications.

In this article, we fabricated a silica/carbonyl iron powder (SiO$_2$@Fe) core-shell structure. In order to study the thermal stability and electromagnetic properties after heat treatment, the silica coated and uncoated samples were annealed at different temperatures in an air atmosphere. Then, a series of measurements were performed, such as crystal structure, hysteresis loop, and microwave permeability. In addition, we also studied the relationship between the mass increase ratio after heat treatment and the high-temperature annealing magnetic properties.

2. Materials and Methods

Raw CIPs were purchased from Jilin Zhuochuang New Materials Co., Ltd., Jilin, China. The powders were washed in acetone at 50 °C with refluxing for 2 h, followed by drying under vacuum at 50 °C for 3 h; 4.6 g TEOS and 2.8 g deionized water were mixed with 40 mL acetone at room temperature by mechanical stirring for half an hour in a flask, and then 100 g washed iron powders were added to the mixed solution, and the stirring speed increased to 200 rpm/min. After 2 h, 1 mL ammonia solution (25%) was dropped into the flask to promote the progression of the hydrolysis reaction. After 24 h, the product was washed with acetone three times on a suction filter and then dried under vacuum at 50 °C for 6 h. For convenience, the raw CIPs and SiO$_2$@Fe powders were named sample A and sample B, respectively

The crystal structures of the samples were analyzed by X-ray diffraction (XRD) on a diffractometer (Philips Panalytical X'pert, Amsterdam, Holland) with Cu Kα radiation. Photos were taken on a field emission scanning electron microscope (Hitachi S-4800, Tokyo, Japan) (SEM) and transmission electron microscope (TEM) (JIM-2010 Hitachi Tokyo Japan). The samples for microwave electromagnetic properties measurements were mixed with paraffin (mass ratio of 15%) and pressed into a ring shape with a 7.00 mm outer diameter and 3.00 mm inner diameter with a thickness of 2 mm. The scattering parameters (S_{11}, S_{21}) were measured by a network analyzer (Agilent Technologies E8363B, Santa Clara CA, United States) in the range of 1~18 GHz. All measurements were performed at room temperature.

3. Results and Discussion

Figure 1a shows the SEM photo of sample A, and the raw carbonyl iron powders were ball-shaped particles with diameters ranging 1~3 μm. Figure 1b shows the TEM image of sample A, and it seems that the particles of sample A had a rough surface. Figure 1c indicates the EDS spectrum of sample A; Fe, C, and O elements. Figure 1e shows the SEM picture of sample B, and the coated particles had a similar shape compared with the raw powders. Figure 1f clearly shows a thin and complete layer on the particle's surface in sample B with a thickness of around 90 nm. Furthermore, the EDS spectrum of sample B indicates the presence of silicon on the surface of particles after coating treatment, as shown in Figure 1g. Figure 1d,f show the particle size distribution calculated from the SEM photos. It can be seen that the particle size of both samples varied from 0.5 to 5.5 μm.

The XRD patterns of sample A and sample B from 10° to 90° are shown in Figure 2. There were three peaks at 2θ equal to 44.6°, 65.1°, and 82.4° in the XRD pattern of sample A, and the peaks can be indexed to the (110), (200), and (211) planes of cubic α-Fe (#87-0731). The pattern of sample B was similar to that of sample A. There was no trace of SiO$_2$ in sample B, suggesting the coating should be an amorphous structure. The FT–IR spectra of sample A and sample B are shown in Figure S1. It was observed that there were three peaks corresponding to the Si-O-Si bands. In summary, all the evidence indicates that the raw carbonyl iron powders were successfully coated with a tight amorphous silica layer. The thermal gravity (TG) curves of sample A and sample B are showed in Figure S2.

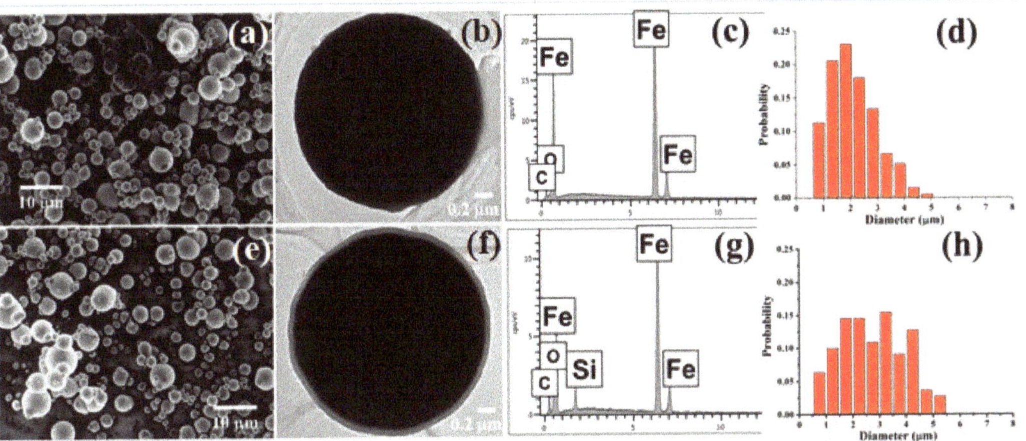

Figure 1. Scanning Electron Microscope (SEM) (**a**) and Transmission Electron Microscope (TEM) (**b**) images of sample A, EDS spectrum of sample A for the selected area (**c**), and particle size distribution of sample A (**d**); SEM (**e**) and TEM (**f**) images of sample B, EDS spectrum of sample B for the selected area (**g**), and particle size distribution of sample B (**h**).

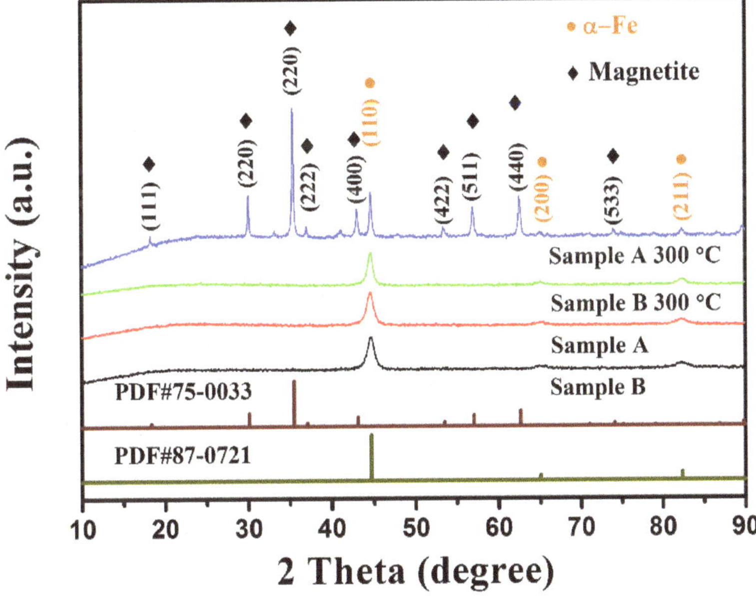

Figure 2. The XRD patterns of samples A and B before and after annealing at 300 °C.

Sample A and sample B were annealed at 220 °C, 250 °C, and 300 °C for 6 h in an air atmosphere and then naturally cooled to room temperature. Sample A annealed at different temperatures was sintered into agglomerates, and the color part turned red, indicating that it was severely oxidized. However, no apparent aggregation was observed for all annealed samples B, and there was almost no change in color. Therefore, the mass ratio of samples A and B before and after annealing was calculated.

Figure 2 shows the XRD spectrum of sample A before and after annealing at 300 °C for 6 h in an air atmosphere. It can be observed that there were two independent phases in the annealed sample A; the primary phase was magnetite (PDF#75-0033), and the rest was α-Fe (PDF#87-0721). After annealing, the corresponding increase in the mass ratio of sample A reached 23% (Table 1), which is lower than the theoretical increase in the mass ratio of α-Fe completely converted into magnetite by 38%. The results show that about 60.8% of the iron atoms in the annealed sample A were oxidized, resulting in a significant change in magnetization. According to the FWHM of the diffraction peak, the grain size of sample A increased significantly.

Table 1. Incremental mass ratio and ratio of oxidized Fe atoms for sample A and sample B after annealing at different temperatures.

Annealing Temperature	Sample A Incremental Mass Ratio	Sample A Ratio of Oxidized Fe Atoms	Sample B Incremental Mass Ratio	Sample B Ratio of Oxidized Fe Atoms
220 °C			0.5%	1.3%
250 °C			2.0%	5.0%
300 °C	23%	60.8%	3.8%	10.0%

Figure 3 shows the hysteresis loop for sample A before and after annealing. It was found that the M_s of annealed sample A drastically dropped sharply from 201.5 to 112.5 emu/g, and the H_c rose from 7.5 to 180 Oe. The decrease in M_s is mainly due to the presence of magnetite with M_s lower than that of α-Fe in annealed sample A. The size distribution of both samples was 0.5 to 5.5 µm, which is much larger than the single-domain critical size of iron particles [17]. Therefore, carbonyl iron powder's internal magnetic domain is a multi-domain state. Under the action of an external magnetic field, the magnetization reversal process is dominated by domain wall motion.

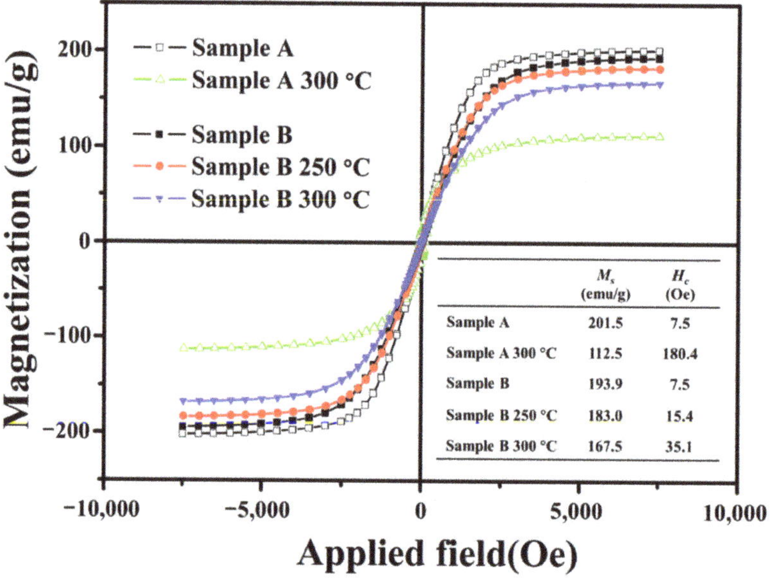

	M_s (emu/g)	H_c (Oe)
Sample A	201.5	7.5
Sample A 300 °C	112.5	180.4
Sample B	193.9	7.5
Sample B 250 °C	183.0	15.4
Sample B 300 °C	167.5	35.1

Figure 3. The hysteresis loops of samples A and B before and after annealing at 250 °C and 300 °C.

According to reports, the M_s of magnetite is about 75 emu/g [18]. Furthermore, the formed magnetite may act as a "pin" preventing the reversal of the magnetization of the whole particle, resulting in an increase in H_c.

Figure 4a shows the real part (ε') and imaginary part (ε'') of the permittivity of sample A before and after annealing at 300 °C from 2 to 18 GHz. It was found that the ε' and ε'' of sample A had a considerable increase after annealing. Figure 4b indicates the real part (μ') and imaginary part (μ'') of the permeability of sample A before and after annealing at 300 °C in the measured range. It was found that the μ' and μ'' of sample B both dropped sharply after annealing within the measured range. There were two peaks in the μ'' spectrum of sample A: the first peak was around 5.5 GHz and the second around 10 GHz, which can be ascribed to domain wall motion at lower frequencies and spin rotation at higher frequencies [19].

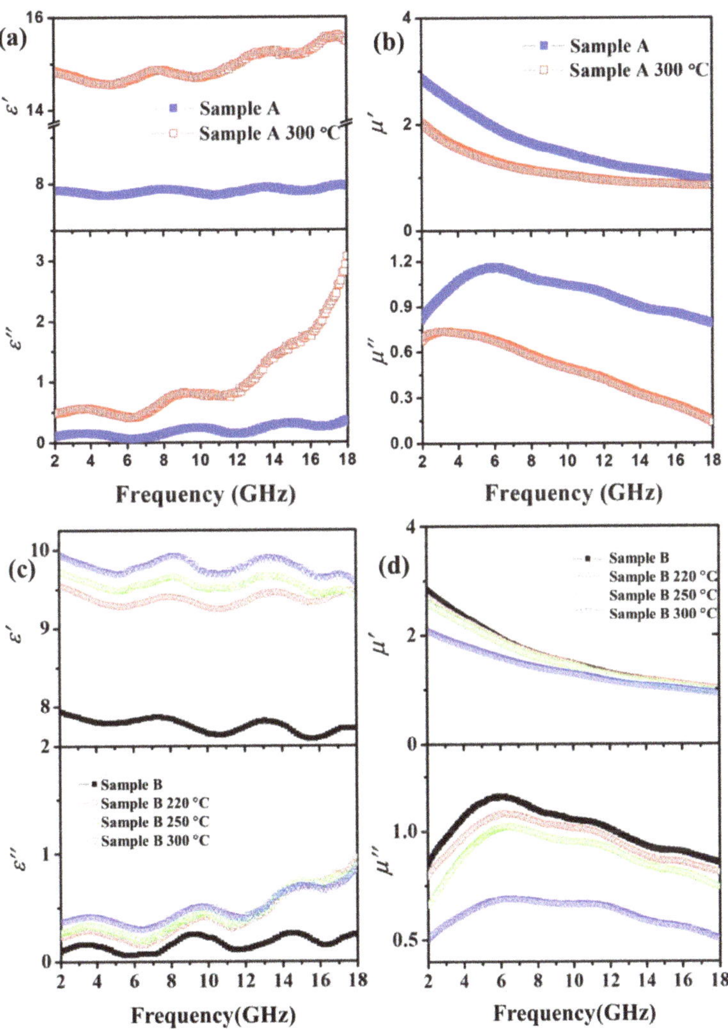

Figure 4. The permittivity spectra (**a**) and permeability spectra (**b**) of sample A before and after annealing at 300 °C; the permittivity spectra (**c**) and permeability spectra (**d**) of sample B before and after annealing at 220 °C, 250 °C, and 300 °C.

According to the transmit-line theory [20], the reflection loss (RL) of sample A before and after annealing at different thicknesses in the range of 2–18 GHz was calculated by the following equations:

$$RL = 20\lg\left|\frac{Z_{in} - Z_0}{Z_{in} + Z_0}\right| \tag{1}$$

$$Z_{in} = Z_0\sqrt{\frac{\mu_r}{\varepsilon_r}}\tanh\left(j\frac{2\pi f}{c}d\sqrt{\mu_r\varepsilon_r}\right) \tag{2}$$

where Z_{in} is the impedance of the incident wave at the interface between the free space and the material, also called the input impedance; Z_0 is the impedance of the incident wave in free space, called the intrinsic impedance; $\mu_r = \mu' - j\mu''$ is the complex permeability, $\varepsilon_r = \varepsilon' - j\varepsilon''$ is the complex permittivity; c is the speed of light in vacuum; d is the thickness of the absorber; and f is the frequency.

It is worth noting that in the first equation, RL can reach the minimum value when $|Z_{in} - Z_0|$ is infinitely close to zero. Therefore, the impedance matching value Z can be defined as [21]:

$$Z = |Z_{in}/Z_0| \tag{3}$$

The closer Z is to 1, the better the impedance matching. Figure 5a,b show the RL map of sample A before and after annealing in the frequency range of 2.0~18.0 GHz with varied absorber thickness from 1.0 to 5.0 mm. It can be observed that the area with qualified microwave absorption (<-10 dB, 90% absorption) in the annealed sample A was significantly suppressed compared to the former. It is widely accepted that the RL intensity and qualified absorption area are the basis for evaluating an eligible absorber [5]. The optimal RL value of sample A was -41.6 dB at 9.7 GHz, and the thickness was 1.875 mm. However, the optimal RL value of the annealed sample A was only -30 dB at 3.8 GHz and the thickness was 3.7 mm. Compared with the RL intensity, the qualified absorption area was more important, because when the RL value was lower than -10 dB, the absorption efficiency was within the acceptable range from 90% to 100% [22]. On the other hand, the bandwidth for RL < -10 dB of sample A was 9.2 GHz, and when the thickness was 1.5 mm, it covered the entire X and Ku bands. However, under the same thickness, the absorption band of the annealed sample A only covered part of the X and Ku bands, and the corresponding bandwidth was reduced to 4.9 GHz. Therefore, it can be confirmed that the microwave absorption capacity of sample A had a considerable decrease after annealing.

After heat treatment, sample A became a composite of magnetite and α-Fe, and the grain size increased. The presence of the magnetite weakened the interaction of the magnetic particles, resulting in a decrease in the values of μ' and μ'' after annealing [5]. In addition, a large number of defects may be generated on the new interface between magnetite and α-Fe in the annealed sample A during the annealing process, which acts as a new polarization center, resulting in a sharp increase in ε' and ε'' [23]. As a result, the gap between the complex permittivity and the magnetic permeability was larger than the former, resulting in a decrease in the impedance matching value Z.

Figure 6a,b show the relationship between Z in sample A and the frequency from 2 GHz to 18 GHz before and after annealing when the absorber thickness was changed from 1 to 5 mm. It was observed that the Z value of the annealed sample A was significantly lower than that of sample B at all thicknesses in the measurement range. The maximum value of Z for sample A was 0.54 at 4.4 GHz, but after annealing, the maximum value of Z decreased to 0.33 at 3.8 GHz. Good microwave absorption can be achieved when Z is close to 1. However, the Z value of the annealed sample A was far from 1, which means that the impedance was not well matched and more microwaves were reflected instead of entering the absorber interface. All of the facts demonstrate the microwave-absorbing ability deteriorated for sample A after annealing, which can be ascribed to the poor impedance matching.

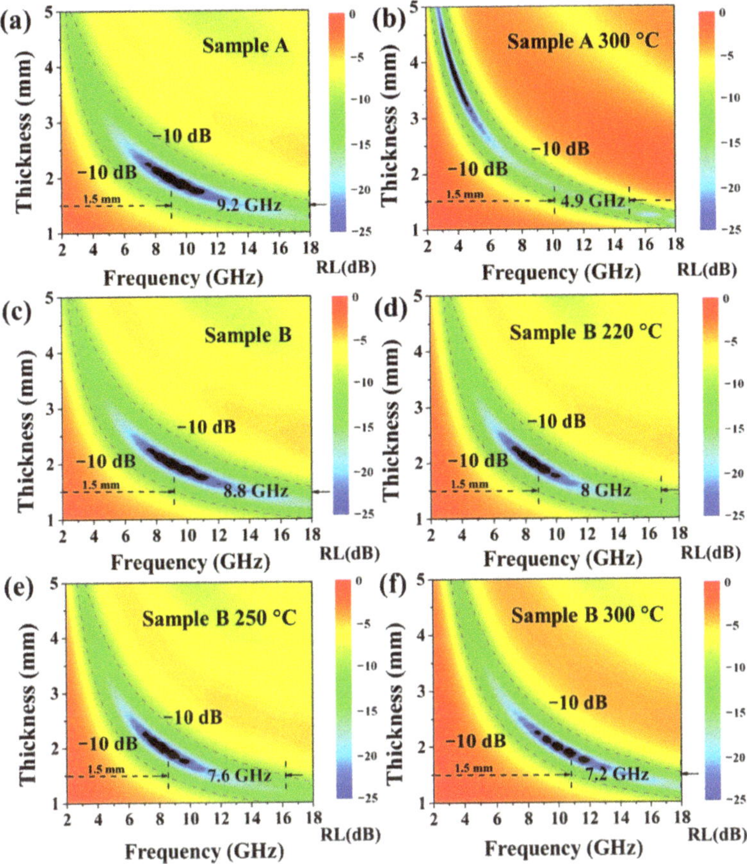

Figure 5. Reflection loss (RL) maps of sample A (**a**), sample A annealing at 300 °C (**b**), sample B (**c**), sample B annealing at 220 °C (**d**), sample B annealing at 250 °C (**e**), and sample B annealing at 300 °C (**f**).

Figure 2 shows the XRD spectra of sample B before and after annealing at 300 °C. It was observed that the spectrum of sample B after annealing maintained the same trend as that of sample B, and there was only one α-Fe phase in the sample. After annealing, the corresponding mass ratio increase in sample B was 3.8%, and the calculated iron atom oxidation ratio was 10% (Table 1), which is much lower than that of the annealed sample A, indicating that oxidation was basically suppressed by the silica coating.

Figure 3 shows the magnetic hysteresis loops of sample B before and after annealing in air atmosphere at 250 and 300 °C. It can be found that the M_s of sample B gradually decreased with the elevated annealing temperature, while H_c increased (insert table in Figure 3). On the other hand, it was noticed that as the annealing temperature increased, the incremental mass ratio of sample B after annealing increased from 0.5% to 3.8% (Table 1), and the calculated ratio of the iron oxide atoms increased from 1.3% to 10.0%. We believe that the incremental mass of sample B after annealing was from the magnetite in it, which was similar to that in the annealed sample A.

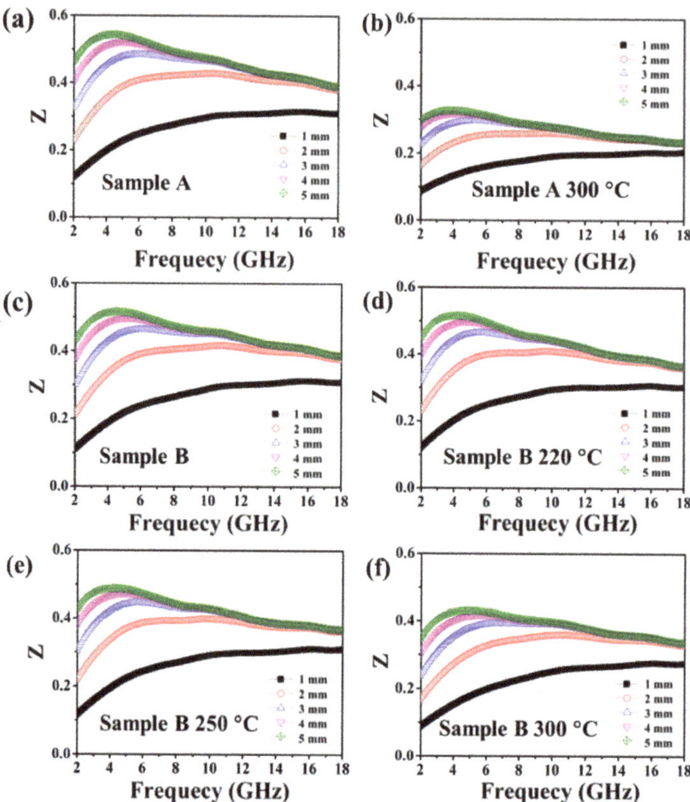

Figure 6. The impedance matching value Z of sample A (**a**), sample A annealing at 300 °C (**b**), sample B (**c**), sample B annealing at 220 °C (**d**), sample B annealing at 250 °C (**e**), and sample B annealing at 300 °C (**f**).

Figure 4c shows the ε' and ε'' of the annealed sample B from 2 to 18 GHz. After annealing, the ε' and ε'' of sample B maintained the same trend in the measurement range compared with that of the un-annealed sample, and the initial value increased with the increase in annealing temperature.

Figure 4d shows the μ' and μ'' spectra of the annealed sample B from 2 to 18 GHz. It can be observed that the permeability spectra of sample B annealed at 220 °C and 250 °C were similar to those of sample B, and the value of μ' and μ'' slightly decreased when the annealing temperature increased. However, when the annealing temperature increased to 300 °C, both the real and imaginary parts of sample B apparently decreased. This means that the silica protective coating started to lose its effectiveness at this temperature. This could be caused by the degradation of SiO_2 coatings, which is determined by delamination and crack formation in a high-temperature environment [24]. Further evidence can be seen in Figures S3 and S4. In terms of the incremental mass ratio after annealing, 2% was a tipping point for permeability reduction.

Figure 5c–f show the RL maps of sample B before and after annealing at 220 °C, 250 °C, and 300 °C, respectively, with varied absorber thickness from 1 to 5 mm in the frequency range 2~18 GHz. The optimal RL of sample B was −41 dB at 9 GHz for a thickness of 2.0 mm. When the annealing temperature was 220 °C, 250 °C, and 300 °C, the optimal RL of the samples was −39.3 dB at 8.4 GHz for a thickness of 2.0 mm, −42.6 dB at 8.2 GHz for a thickness of 2.0 mm, and −39.6 dB at 10.6 GHz for a thickness of 1.9 mm, respectively. It was also observed that the bandwidth for RL<–10 dB of sample B decreased from 8.8 GHz to 7.2 GHz with increased annealing temperature when the absorber thickness was 1.5 mm. From this point of view, the microwave absorption performance of sample B was not significantly affected after annealing.

Figure 6c–f indicate the impedance matching value Z of sample B (a) and sample B annealing at 220 °C (b), 250 °C (b), and 300 °C (d) with varied absorber thickness from 1 to 5 mm in the frequency range from 2 to 18 GHz. For all samples, the highest Z value occurred when the thickness was 5 mm. The maximum Z value of sample B was 0.52. For the annealed samples, the maximum values were 0.51, 0.49, and 0.43 when the annealing temperature was 220 °C, 250 °C, and 300 °C, respectively. It was observed that when the annealing temperature increased, the Z value decreased slowly, which means that sample B maintained a good impedance match compared with sample A.

As described above, the gap between the complex permittivity and magnetic permeability of sample A after annealing treatment increased, resulting in poor impedance matching. In addition, we noticed that the changes in the complex permittivity of samples A and B during the heat treatment were greater than the changes in magnetic permeability. It is speculated that dielectric loss may play an important role in microwave absorption. The dielectric loss process can be explained by the Debye theory [25–27]; the relationship between ε' and ε'' is written as:

$$\left(\varepsilon' - \frac{(\varepsilon_s + \varepsilon_\infty)}{2}\right)^2 + (\varepsilon'')^2 = \frac{(\varepsilon_s + \varepsilon_\infty)^2}{2} \tag{4}$$

where ε_s is the static permittivity, and ε_∞ is relative permittivity at infinite frequency. The shape of the curve is a semicircle, called to a Cole–Cole semicircle [28,29]. In detail, enhancement of the Debye dipolar relaxation is accompanied by an increased number of semicircles, an expanded semicircle radius, and a higher frequency position. Figure 7 shows the relationship of sample A and B between ε' and ε'' before and after annealing.

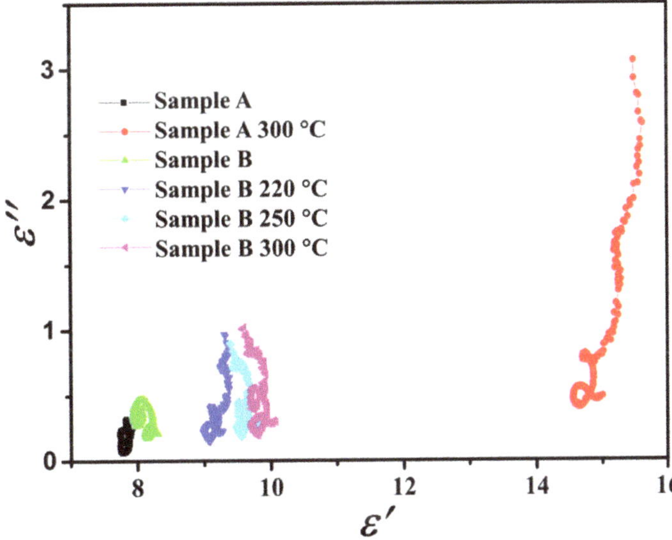

Figure 7. Typical Cole–Cole semicircles of ε' vs. ε'' for sample A and B before and after annealing.

It was clearly observed that the Cole–Cole semicircles of the annealed sample A had the largest radius and highest frequency position, indicating the strongest Debye dipolar relaxation. The strongest Debye relaxation not only brings about supreme dielectric loss, but also causes improper impedance matching.

4. Conclusions

Silica-coated CIPs with a core-shell structure were fabricated and annealed at elevated temperature in an air atmosphere to investigate the thermal stability and the variation of magnetic properties. It was found that the coated CIP maintained good microwave absorption performance after annealing compared to the uncoated one. It was confirmed that the silica layer significantly prevented the oxidation of the CIP, which leads to poor impedance matching and microwave absorption performance. It is worth mentioning that the mass increase ratio of the sample after annealing was a reference value for studying the change in its magnetic properties. With the increase in the mass ratio after annealing, the M_s decreased and H_c increased, accompanied by the rise in the ε' and ε'' and a reduction in the μ' and μ''. It is a reliable and straightforward method to measure the oxidation resistance of coated iron powder by calculating the incremental mass ratio of samples before and after annealing. In this article, an incremental mass ratio less than 2% indicated that the electromagnetic properties of the sample had not changed much compared to those before annealing and had a good impedance match. This article provides a new perspective for studying the high-temperature resistance of soft magnetic materials.

Supplementary Materials: The following supporting information can be downloaded at: https://www.mdpi.com/article/10.3390/ma15072499/s1, Figure S1: Fourier transform infrared spectroscopy (FT-IR) spectra of Sample A and Sample B; Figure S2: The thermogravimetric (TG) curves of Sample A and Sample B in air atmosphere; Figure S3: The Transmission Electron Microscope (TEM) photos of Sample B annealing at 300 °C; Figure S4: The X-ray photoelectron spectroscopy (XPS) spectra of Sample B, Sample B annealing at 250 °C, and Sample B annealing at 300 °C.

Author Contributions: Conceptualization, X.Y., X.M. and B.H.; Investigation, X.Y., X.M. and Q.Z.; Supervision, B.H.; Visualization, X.Y., C.S. and Z.M.; Writing—original draft, X.Y.; Writing—review & editing, C.S., Q.Z. and Z.M.; Funding acquisition, X.Y. and Z.M. All authors have read and agreed to the published version of the manuscript.

Funding: This research was financially supported by the Special project of high tech industrialization of science and technology cooperation between Jilin Province and the Chinese Academy of Sciences (Grant No. 2021SYHZ0037) and the CAS "Light of West China" Program and the Science and Technology Plan of Gansu Province (No. 20JR10RA044).

Institutional Review Board Statement: Not applicable.

Informed Consent Statement: Not applicable.

Data Availability Statement: Not applicable.

Acknowledgments: Among the authors, Mu offers a solution to the coating device design and many other useful suggestions, with equal contributions as the first author.

Conflicts of Interest: The authors declare no conflict of interest.

References

1. Park, K.Y.; Han, J.H.; Lee, S.B.; Yi, J.W. Microwave absorbing hybrid composites containing Ni–Fe coated carbon nanofibers prepared by electroless plating. *Compos. Part A Appl. Sci. Manuf.* **2011**, *42*, 573–578. [CrossRef]
2. Zhang, L.; Shi, C.S.; Rhee, K.Y.; Zhao, N.Q. Properties of $Co_{0.5}Ni_{0.5}Fe_2O_4$/carbon nanotubes/polyimide nanocomposites for microwave absorption. *Compos. Part A Appl. Sci. Manuf.* **2012**, *43*, 2241–2248. [CrossRef]
3. Niu, D.Y.; Yu, J.K.; Qiao, Q.D.; Wang, D.; Liu, R.H. Improved corrosion resisting property of magnetism iron fiber by SiO_2 coating. *J. Surf. Eng. Mater. Adv. Technol.* **2012**, *2*, 163–166.
4. Ramo, S.; Whinnery, J.R.; Duzer, T.V. *Fields and Waves in Communication*, 3rd ed.; Electronics Wiley: New York, NY, USA, 1994; pp. 664–665.
5. Duan, W.J.; Li, X.D.; Wang, Y.; Qiang, R.; Tian, C.H.; Wang, N.; Han, X.J.; Du, Y.C. Surface functionalization of carbonyl iron with aluminum phosphate coating toward enhanced anti-oxidative ability and microwave absorption properties. *Appl. Surf. Sci.* **2018**, *427*, 594–602. [CrossRef]
6. Abshinova, M.A.; Kazantseva, N.E.; Sáha, P.; Sapurina, I.; Kovářová, J.; Stejskal, J. The enhancement of the oxidation resistance of carbonyl iron by polyaniline coating and consequent changes in electromagnetic properties. *Polym. Degrad. Stab.* **2008**, *93*, 1826–1831. [CrossRef]

7. Bruncková, H.; Kabátová, M.; Dudrová, E. The effect of iron phosphate, alumina and silica coatings on the morphology of carbonyl iron particles. *Surf. Interface Anal.* **2010**, *42*, 13–20. [CrossRef]

8. Jaya, F.; Gauthier-Brunet, V.; Pailloux, F.; Mimault, J.; Bucher, S.; Dubois, S. Al-coated iron particles: Synthesis, characterization and improvement of oxidation resistance. *Surf. Coat. Technol.* **2008**, *202*, 4302–4306. [CrossRef]

9. Zhou, Y.Y.; Ma, L.Y.; Li, R.; Chen, D.; Lu, Y.Y.; Cheng, Y.Y.; Luo, X.X.; Xie, H.; Zhou, W.C. Enhanced heat-resistance property of aluminum-coated carbonyl iron particles as microwave absorption materials. *J. Magn. Magn. Mater.* **2021**, *524*, 167681. [CrossRef]

10. Xie, H.; Zhou, Y.Y.; Ren, Z.W.; Wei, X.; Tao, S.P.; Yang, C.Q. Enhancement of electromagnetic interference shielding and heat-resistance properties of silver-coated carbonyl iron powders composite material. *J. Magn. Magn. Mater.* **2020**, *499*, 166244. [CrossRef]

11. Zhou, Y.Y.; Zhou, W.C.; Li, R.; Mu, Y.; Qing, Y.C. Enhanced antioxidation and electromagnetic properties of Co-coated flaky carbonyl iron particles prepared by electroless plating. *J. Alloy. Compd.* **2015**, *637*, 10–15. [CrossRef]

12. Jia, S.; Luo, F.; Qing, Y.C.; Zhou, W.C.; Zhu, D.M. Electroless plating preparation and microwave electromagnetic properties of Ni-coated carbonyl iron particle/epoxy coatings. *Physica B* **2010**, *405*, 3611–3615. [CrossRef]

13. Zhou, Y.Y.; Xie, H.; Zhou, W.C.; Ren, Z.W. Enhanced antioxidation and microwave absorbing properties of SiO_2-coated flaky carbonyl iron particles. *J. Magn. Magn. Mater.* **2018**, *446*, 143–149. [CrossRef]

14. Zhu, J.H.; Wei, S.Y.; Lee, I.Y.; Park, S.; Willis, J.; Haldolaarachchige, N.; Young, D.P.; Luo, Z.P.; Guo, Z.H. Silica stabilized iron particles toward anti-corrosion magnetic polyurethane nanocomposites. *RSC Adv.* **2012**, *2*, 1136–1143. [CrossRef]

15. Qing, Y.C.; Zhou, W.C.; Jia, S.; Luo, F.; Zhu, D.M. Microwave electromagnetic property of SiO_2-coated carbonyl iron particles with higher oxidation resistance. *Physica B* **2011**, *406*, 777–780.

16. Maklakov, S.S.; Lagarkov, A.N.; Maklakov, S.A.; Adamovich, Y.A.; Petrov, D.A.; Rozanov, K.N.; Ryzhikov, I.A.; Zarubina, A.Y.; Pokholok, K.V.; Filimonov, D.S. Corrosion-resistive magnetic powder $Fe@SiO_2$ for microwave applications. *J. Alloy. Compd.* **2017**, *706*, 267–273. [CrossRef]

17. Butler, R.F.; Banerjee, S.K. Single-domain grain size limits for metallic iron. *J. Geophys. Res.* **1975**, *80*, 252. [CrossRef]

18. Goya, G.F.; Berquó, T.S.; Fonseca, F.C.; Morales, M.P. Static and dynamic magnetic properties of spherical magnetite nanoparticles. *J. Appl. Phys.* **2003**, *94*, 3520. [CrossRef]

19. Qing, Y.C.; Zhou, W.C.; Luo, F.; Zhu, D.M. Epoxy-silicone filled with multi-walled carbon nanotubes and carbonyl iron particles as a microwave absorber. *Carbon* **2010**, *48*, 4074. [CrossRef]

20. Naito, Y.; Suetake, K. Application of Ferrite to Electromagnetic Wave Absorber and its Characteristics. *IEEE Trans. Microw. Theory* **1971**, *19*, 65–72. [CrossRef]

21. Zhang, Y.; Wen, J.; Zhang, L.; Lu, H.; Guo, Y.; Ma, X.; Zhang, M.; Yin, J.; Dai, L.; Jian, X.; et al. High antioxidant lamellar structure Cr_2AlC: Dielectric and microwave absorption properties in X band. *J. Alloy. Compd.* **2021**, *860*, 157896. [CrossRef]

22. Wang, Y.; Du, Y.; Qiang, R.; Tian, C.; Xu, P.; Han, X. Interfacially engineered sandwich-like rGO/carbon microspheres/rGO composite as an efficient and durable microwave absorber. *Adv. Mater. Interfaces* **2016**, *3*, 1500684. [CrossRef]

23. Jiang, X.Y.; Wan, W.H.; Wang, B.; Zhang, L.B.; Yin, L.J.; Bui, H.V.; Xie, J.L.; Zhang, L.; Lu, H.P.; Deng, L.J. Enhanced anti-corrosion and microwave absorption performance with carbonyl iron modified by organic fluorinated chemicals. *Appl. Surf. Sci.* **2022**, *572*, 151320. [CrossRef]

24. Hofman, R.; Westheim, J.G.F.; Pouwel, I.; Fransen, T.; Gellings, P.J. FTIR and XPS Studies on Corrosion-resistant SiO_2 Coatings as a Function of the Humidity during Deposition. *Surf. Interface Anal.* **1996**, *24*, 1–6. [CrossRef]

25. Yu, H.; Wang, T.; Wen, B.; Lu, M.; Xu, Z.; Zhu, C.; Chen, Y.; Xue, X.; Sun, C.; Cao, M. Graphene/polyaniline nanorod arrays: Synthesis and excellent electromagnetic absorption properties. *J. Mater. Chem.* **2012**, *22*, 21679–21685. [CrossRef]

26. Meng, F.B.; Wang, H.G.; Huang, F.; Guo, Y.F.; Wang, Z.Y.; Hui, D.; Zhou, Z.W. Graphene-based microwave absorbing composites: A review and prospective. *Compos. B Eng.* **2018**, *137*, 260–277. [CrossRef]

27. Li, Q.S.; Zhu, J.J.; Wang, S.N.; Huang, F.; Liu, Q.C.; Kong, X.K. Microwave absorption on a bare biomass derived holey silica-hybridized carbon absorbent. *Carbon* **2020**, *161*, 639–646. [CrossRef]

28. Wang, X.; Pan, F.; Xiang, Z.; Zeng, Q.W.; Pei, K.; Che, R.C.; Lu, W. Magnetic vortex coreshell $Fe_3O_4@C$ nanorings with enhanced microwave absorption performance. *Carbon* **2020**, *157*, 130–139. [CrossRef]

29. Li, Q.S.; Li, S.T.; Liu, Q.G.; Liu, X.F.; Shui, J.L.; Kong, X.K. Iodine cation bridged graphene sheets with strengthened interface combination for electromagnetic wave absorption. *Carbon* **2021**, *183*, 100–107. [CrossRef]

MDPI

St. Alban-Anlage 66

4052 Basel

Switzerland

Tel. +41 61 683 77 34

Fax +41 61 302 89 18

www.mdpi.com

Materials Editorial Office

E-mail: materials@mdpi.com

www.mdpi.com/journal/materials